The AMAZING ALGEBRA BOOK

Grades 6–12

20 ENGAGING TRICKS

Julian F. Fleron and Ronald Edwards

Printed in the United States of America.

This book is printed on recycled paper.

Order Number 2-5277
ISBN 978-1-58324-259-9

E F G H I 16 15 14 13 12

395 Main Street
Rowley, MA 01969
www.didax.com

Contents

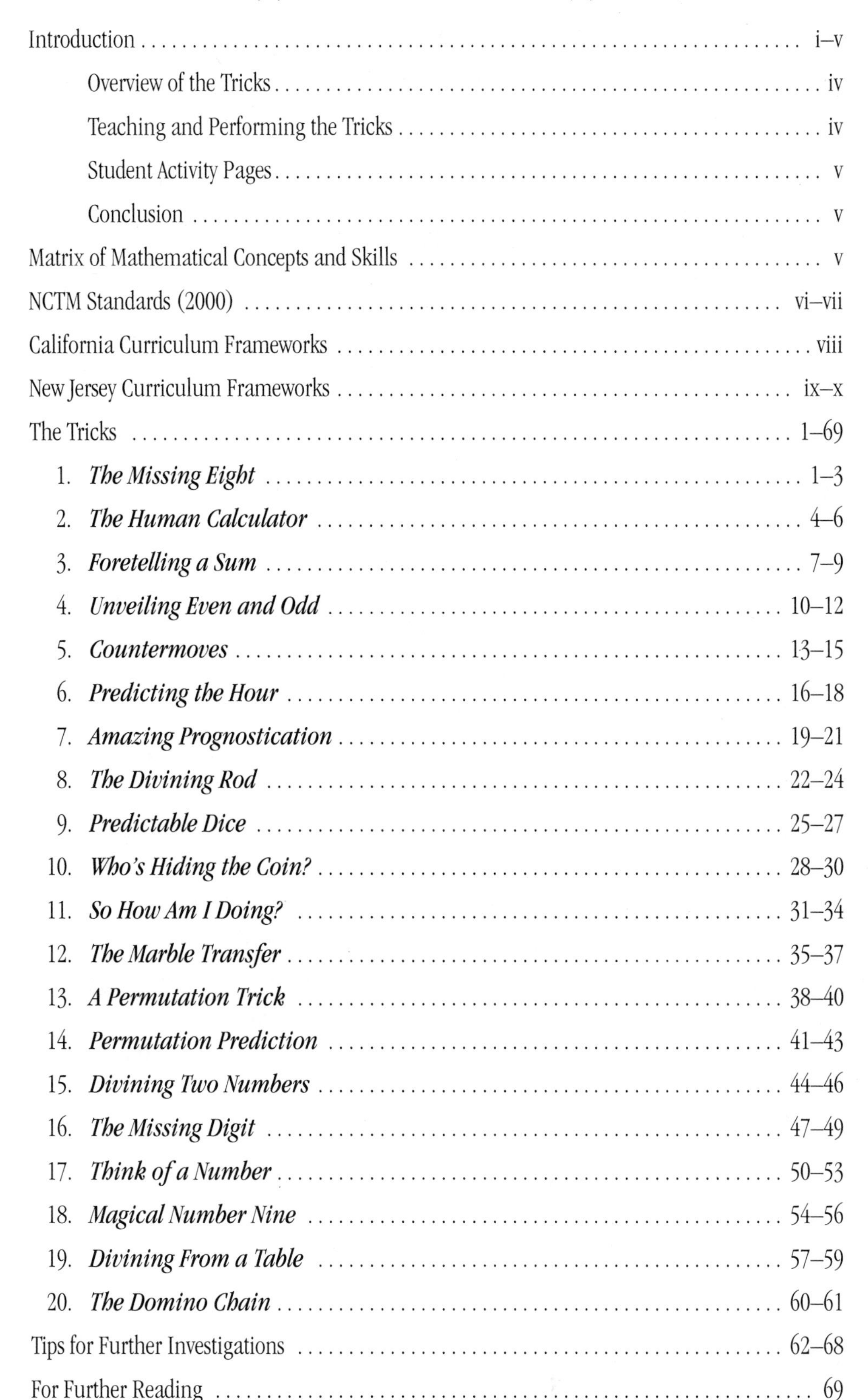

Introduction

The goal of ***The Amazing Algebra Book*** is to provide a repertoire of mathematical magic tricks and accompanying classroom materials for developing skills and an understanding of algebra. While seeking to nurture and motivate algebra students, these tricks allow students to make substantive progress in meeting required national and state curricular requirements. The book contains twenty magic tricks, some new, some classical, which will amaze and delight your students. They will enrich, captivate, and motivate pre-algebra and algebra students of all ages. But the goal is not simply to amaze. Underlying the mystery of each trick are the central tools and concepts in beginning algebra which appear in matrices of skills and concepts keyed to both national standards and key state curriculum frameworks.

The Amazing Algebra Book has been developed for ease of use by classroom teachers. The tricks themselves, the curricular connections, and the mathematical details are clearly spelled out in the *Teacher Notes*. *Student Activity Pages* are ready to copy. These materials are easily incorporated into typical algebra classrooms, but retain enough flexibility for a broad range of classroom environments and a variety of teaching and learning styles. First-year algebra students should be capable of understanding all algebraic arguments, which explain the mathematics behind the tricks. These tricks also provide appropriate experiences for *math clubs*, *math teams*, *gifted and talented programs*, *after-school enrichment programs*, or students who are *home-schooled*.

Each trick is organized into three sections: *Overview of the Trick*, *Teaching and Performing the Trick*, and a *Student Activity Page*.

Overview of the Tricks

This section gives a brief synopsis of the trick to give the teacher a sense of what it entails. It also presents a concrete example of the trick and a brief explanation of how the trick works. Once you have read the *Overview of the Trick*, you will have a clear understanding of the trick and the algebra skills involved. Then you can decide how it will fit into your curriculum. This section has a list of the main skills and concepts which will insure students are adequately prepared to investigate the trick. These skills and concepts are keyed to corresponding matrices of skills and concepts, which appear after the *Introduction*.

Teaching and Performing the Tricks

This section addresses possible modes of instruction, classroom materials needed, the performance of the trick, and an algebraic analysis of why the trick works.

Modes of Instruction: We expect you will tailor instruction to your classroom's needs and your teaching style, but we hope that by illustrating a variety of different approaches you may experiment with new modes of instruction. Below is a brief description of different instructional modes and a list of tricks that illustrate the different modes.

Direct teaching

The teacher directs the performance of the trick. Student volunteers are selected to help with certain tasks while the class checks their work. (Examples: Tricks 2, 3, 6, 12.)

Whole class

The teacher directs the performance of the trick. However, every student in the class is an active volunteer. (Examples: Tricks 4, 8, 11, 16, 18.)

Small groups

Student apprentices perform the tricks in small groups while the teacher monitors students' progress. (Examples: Tricks 1, 7, 14, 17, 19.)

Cooperative groups

Students participate in the trick working cooperatively in groups with all of the students active. (Examples: Tricks 5, 9, 13, 20.)

Discovery learning

Students actively discover the algebraic mechanisms that underlie the trick. This allows students to discover the trick on their own and helps them become independent learners. (Examples: Tricks 10, 15.)

Materials: The majority of the tricks require little more than chalk/whiteboards, paper and pencils, and optional use of calculators. There are a few additional materials that are necessary for specific tricks. Most of these materials are standard mathematical manipulatives or are easily constructible.

Performing the Trick: In this section, there are step-by-step instructions on how to perform the trick in a classroom setting. As noted above, the mode of instruction used to illustrate this section is varied from trick to trick. This is done in the hope that you might explore modes of instruction you might not regularly employ.

Analysis – Using algebra to explain how it works: This section presents an algebraic analysis of the trick. Complete details are provided to explain the mechanisms that enable the trick to work. Although the tricks can be analyzed in a number of different ways, the methods illustrated seem to be the clearest and offer the best use of algebraic techniques used in a standard curriculum.

Student Activity Pages

The *Student Activity Pages* were designed to be photocopied and distributed to your students. The *Student Activity Pages* consist of four sections. The first is a brief review of the trick. This is followed by "Unraveling the Trick," which presents a series of leading questions designed to help students discover the algebraic mechanisms upon which the trick is based. The third section, "Extending Questions," brings up important issues in the algebraic analysis of the trick. The *Student Activity Pages* conclude with "Further Investigations." These investigations offer students opportunities to extend the trick, invent related magic tricks, or discover connections with other mathematical topics. The teacher may wish to use these investigations for classroom discussion, projects for group work, student challenge problems, or projects for a math lab or club. Hints to solutions are given at the end of the teacher notes.

Conclusion

As students become more confident in analyzing these tricks, they will be able to design their own magic tricks based on algebra. This last step completes the cycle in a mathematical exploration: I observe, I conjecture, I form a hypothesis, I construct a proof, I create my own magic tricks. At this point, algebra is no longer simply used to describe and illustrate how a specific trick works, but as a tool in creating new ones—students have become original algebraic thinkers. We hope these materials help bring magic into your algebra classroom.

Matrix of Mathematical Concepts

Arithmetic	**1**	**2**	**3**	**4**	**5**	**6**	**7**	**8**	**9**	**10**	**11**	**12**	**13**	**14**	**15**	**16**	**17**	**18**	**19**	**20**
Basic computation	X	X	X	X	X	X	X	X	X	X	X	X	X	X	X	X	X	X	X	
Square roots																	X			
Expanded notation								X			X		X	X	X	X	X		X	
Multiples	X				X						X	X				X	X		X	
Factorization								X			X		X							
Division theorem											X									
Odd and even numbers				X						X				X						X
Prime numbers				X																
Divisibility properties								X			X		X			X	X			
Associative property	X	X																		
Commutative property		X													X					
Distributive property		X				X				X	X				X			X		
Permutations								X					X	X		X				

Algebra	**1**	**2**	**3**	**4**	**5**	**6**	**7**	**8**	**9**	**10**	**11**	**12**	**13**	**14**	**15**	**16**	**17**	**18**	**19**	**20**
Use of variable(s)	X	X	X	X	X	X	X	X	X	X	X	X	X	X	X	X	X	X	X	X
Algebraic expressions	X	X	X	X	X	X	X	X	X	X	X	X	X	X	X	X	X	X		X
Polynomials		X	X	X	X	X	X	X	X	X	X	X	X	X	X	X	X	X		
Simplifying polynomials		X	X		X	X	X	X	X	X	X	X	X	X	X	X	X	X		
Adding polynomials			X		X		X	X	X		X	X	X	X	X	X	X	X		
Multiplying polynomials									X								X			
Distributive property		X	X		X	X	X	X	X	X	X	X	X	X	X	X	X	X		
Factoring polynomials								X			X		X	X		X	X	X		
Solving equations										X		X					X			

NCTM Standards (2000)

***Algebra standard:** Represent and analyze mathematical situations and structures using algebraic symbols.*

Expectations	1	2	3	4	5	6	7	8	9	10	11	12	13	14	15	16	17	18	19	20
Represent the idea of a variable as an unknown quantity using a letter or symbol. (3–5)	X	X	X	X	X	X	X	X	X	X	X	X	X	X	X	X	X	X	X	X
Develop an initial conceptual understanding of different uses of variables. (6–8)	X	X	X	X	X	X	X	X	X	X	X	X	X	X	X	X	X	X	X	X
Identify such properties as commutativity, associativity, and distributivity and use them to compute with whole numbers. (3–5)	X	X	X		X	X	X	X	X	X	X	X	X	X	X	X	X	X		
Express mathematical relationships using equations. (3–5)	X	X	X	X	X	X	X	X	X	X	X	X	X	X	X	X	X	X		X
Use symbolic algebra to represent and explain mathematical relationships. (9–12)	X	X	X	X	X	X	X	X	X	X	X	X	X	X	X	X	X	X		X
Recognize and generate equivalent forms for simple algebraic expressions and solve linear equations. (6–8)	X	X	X		X	X	X	X	X	X	X	X	X	X	X	X	X	X		
Understand the meaning of equivalent forms of expressions, equations, inequalities, and relations. (9–12)	X	X	X		X	X	X	X	X	X	X	X	X	X	X	X	X	X		
Write equivalent forms of equations … and solve them with fluency …. (9–12)	X	X	X		X	X	X	X	X	X	X	X	X	X	X	X	X	X		

***Algebra standard:** Use mathematical models to represent and understand quantitative relationships.*

Expectations	1	2	3	4	5	6	7	8	9	10	11	12	13	14	15	16	17	18	19	20
Model problem situations … and use representations such as … equations to draw conclusions. (3–5)	X	X	X	X	X	X	X	X	X	X	X	X	X	X	X	X	X	X		X
Model and solve contextualized problems using various representations, such as … equations. (6–8)	X	X	X	X	X	X	X	X	X	X	X	X	X	X	X	X	X	X		X
Identify essential quantitative relationships in a situation and determine the class or classes of functions that might model the relationships. (9–12)	X	X	X	X	X	X	X	X	X	X	X	X	X	X	X	X	X	X		X
Use symbolic expressions … to represent relationships arising from various contexts. (9–12)	X	X	X	X	X	X	X	X	X	X	X	X	X	X	X	X	X	X		
Draw reasonable conclusions about a situation being modeled. (9–12)	X	X	X	X	X	X	X	X	X	X	X	X	X	X	X	X	X	X		

***Number and operations standard:** Compute fluently and make reasonable estimates.*

Expectations	1	2	3	4	5	6	7	8	9	10	11	12	13	14	15	16	17	18	19	20
Develop fluency in adding, subtracting, multiplying, and dividing whole numbers and decimals. (3–5)	X	X	X	X	X	X	X	X	X	X	X	X	X	X	X	X	X	X		X
Develop fluency with basic number combinations for multiplication and division and use these combinations to mentally compute related problems such as 30 x 50. (3–5)	X	X	X	X	X	X	X	X	X	X	X	X	X	X	X	X	X	X		X
Select appropriate methods and tools for computing with whole numbers from among mental computation, estimation, calculators, and paper and pencil according to the context and nature of the computation and use the selected method or tool. (3–5)	X	X	X	X	X	X	X	X	X	X	X	X	X	X	X	X	X	X		X

Number and operations standard: Understand numbers, ways of representing numbers, relationships among numbers and number systems.

Expectations	1	2	3	4	5	6	7	8	9	10	11	12	13	14	15	16	17	18	19	20
Understand the place-value structure of the base-10 number system and be able to represent and compare whole numbers and decimals. (3–5)	X	X					X	X		X	X		X	X	X	X		X	X	
Recognize equivalent representations for the same number and generate them by decomposing and composing numbers. (3–5)	X	X						X		X			X	X	X	X		X	X	
Describe classes of numbers according to characteristics such as the nature of their factors. (3–5)				X				X		X			X					X	X	
Use factors, multiples, prime factorization, and relatively prime numbers to solve problems. (6–8)	X			X	X			X		X		X	X			X		X		
Develop meaning for integers and represent and compare quantities with them. (6–8)		X		X			X	X		X						X				
Develop a deeper understanding of very large and very small numbers and of various representations of them. (9–12)	X																			
Use number-theory arguments to justify relationships involving whole numbers. (0–12)	X	X		X			X	X		X		X				X			X	

Number and operations standard: Understand meanings of operations and how they relate to one another.

Expectations	1	2	3	4	5	6	7	8	9	10	11	12	13	14	15	16	17	18	19	20
Understand the effects of multiplying and dividing whole numbers. (3–5)	X	X						X		X	X				X	X				
Identify and use relationships between operations, such as division as the inverse of multiplication, to solve problems. (3–5)						X	X					X					X			
Understand and use the inverse relationships of addition and subtraction, multiplication and division, and squaring and square roots to simplify computations and solve problems. (6–8)						X	X										X			
Understand and use properties of operations, such as the distributivity of multiplication over addition. (3–5)	X	X	X		X	X	X	X	X	X	X	X		X	X	X	X	X		
Use the associative and commutative properties of addition and multiplication and the distributive property of multiplication over addition to simplify computations with integers, fractions, and decimals. (6–8)	X	X	X		X	X	X	X	X	X	X	X	X	X	X	X	X	X		
Understand the meaning and effects of arithmetic operations with … integers. (6–8)	X	X	X	X	X	X	X	X	X	X	X	X	X	X	X	X	X	X		

California Curriculum Frameworks

Goals in mathematics education	1	2	3	4	5	6	7	8	9	10	11	12	13	14	15	16	17	18	19	20
Develop fluency in computational skills.	X	X	X	X	X	X	X	X	X	X	X	X	X	X	X	X	X	X	X	X
Develop an understanding of concepts.	X	X	X	X	X	X	X	X	X	X	X	X	X	X	X	X	X	X	X	X
Become problem-solvers.	X	X	X	X	X	X	X	X	X	X	X	X	X	X	X	X	X	X	X	X
Communicate precisely.	X	X	X	X	X	X	X	X	X	X	X	X	X	X	X	X	X	X	X	X
Reason mathematically.	X	X	X	X	X	X	X	X	X	X	X	X	X	X	X	X	X	X	X	X
Make mathematical connections.	X	X	X	X	X	X	X	X	X	X	X	X	X	X	X	X	X	X	X	X

Expectations	1	2	3	4	5	6	7	8	9	10	11	12	13	14	15	16	17	18	19	20
Use standard notation.	X	X	X	X	X	X	X	X	X	X	X	X	X	X	X	X	X	X	X	
Know and use properties of numbers.	X	X	X	X		X	X	X	X	X	X		X	X	X	X	X	X	X	X
Perform basic arithmetic calculations.	X	X	X	X	X	X	X	X	X	X	X	X	X	X	X	X	X	X	X	X
Determine factors and multiples.	X	X	X	X	X			X			X	X	X			X	X	X		
Solve simple problems.	X	X	X	X	X	X	X	X	X	X	X	X	X	X	X	X	X	X	X	X
Apply strategies for problem-solving.	X	X	X	X	X	X	X	X	X	X	X	X	X	X	X	X	X	X	X	X
Support solutions verbally and with symbolic work.	X	X	X	X	X	X	X	X	X	X	X	X	X	X	X	X	X	X	X	X

Algebra and functions	1	2	3	4	5	6	7	8	9	10	11	12	13	14	15	16	17	18	19	20
Express relationships using algebraic terminology, expressions, and inequalities.	X	X	X	X	X	X	X	X	X	X	X	X	X	X	X	X	X	X		X
Evaluate algebraic expressions.					X					X	X	X	X	X			X	X		
Simplify algebraic expressions.	X	X	X		X	X	X	X	X	X	X	X	X	X	X	X	X	X		
Solve algebraic equations.												X					X			
Apply basic factoring techniques.			X					X			X		X	X	X	X	X			
Determine the domain of independent variables.	X	X	X		X	X	X	X	X	X	X	X	X	X	X	X	X	X		
Solve multistep problems.	X	X	X	X	X	X	X	X	X	X	X	X	X	X	X	X	X	X	X	
Use tables and rules to solve problems.				X	X			X	X	X	X	X		X		X		X	X	
Explain mathematical reasoning.	X	X	X	X	X	X	X	X	X	X	X	X	X	X	X	X	X	X	X	X
Construct valid arguments.	X	X	X	X	X	X	X	X	X	X	X	X	X	X	X	X	X	X	X	X
Generalize to other situations.	X	X	X	X	X	X	X	X	X	X	X	X	X	X	X	X	X	X	X	X

Responsibilities of students	1	2	3	4	5	6	7	8	9	10	11	12	13	14	15	16	17	18	19	20
Learn basic skills.	X	X	X	X	X	X	X	X	X	X	X	X	X	X	X	X	X	X	X	
Reason mathematically.	X	X	X	X		X	X	X	X	X	X		X	X	X	X	X	X	X	X
Employ a variety of problems solving methods.	X	X	X	X	X	X	X	X	X	X	X	X	X	X	X	X	X	X	X	X
Communicate and validate solutions.	X	X	X	X	X			X			X	X	X			X	X	X		
Give accurate and detailed proofs.	X	X	X	X	X	X	X	X	X	X	X	X	X	X	X	X	X	X	X	X

All of the tricks and the accompanying materials have been developed to help teachers meet the *Responsibilities of teachers*:

- Engage and support all students in learning
- Create and maintain effective learning environments
- Organize subject matter for student learning
- Planning instruction and learning experiences for all students
- Assessing student learning

New Jersey Curriculum Frameworks

	1	2	3	4	5	6	7	8	9	10	11	12	13	14	15	16	17	18	19	20
Standard 1: Problem solving	X	X	X	X	X	X	X	X	X	X	X	X	X	X	X	X	X	X	X	X

	1	2	3	4	5	6	7	8	9	10	11	12	13	14	15	16	17	18	19	20
Standard 2: Communication	X	X	X	X	X	X	X	X	X	X	X	X	X	X	X	X	X	X	X	X

	1	2	3	4	5	6	7	8	9	10	11	12	13	14	15	16	17	18	19	20
Standard 3: Connections	X	X	X	X	X	X	X	X	X	X	X	X	X	X	X	X	X	X	X	X

	1	2	3	4	5	6	7	8	9	10	11	12	13	14	15	16	17	18	19	20
Standard 4: Reasoning	X	X	X	X	X	X	X	X	X	X	X	X	X	X	X	X	X	X	X	X

Standard 6: Number sense	1	2	3	4	5	6	7	8	9	10	11	12	13	14	15	16	17	18	19	20
Standard notation.	X	X	X	X	X	X	X	X	X	X	X	X	X	X	X	X	X	X	X	X
Number patterns.	X	X	X	X	X			X	X		X				X	X		X	X	X
Number theory concepts.	X	X		X				X			X					X		X	X	
Number properties.	X	X		X	X	X	X	X	X	X	X	X	X	X	X	X	X	X	X	X
Informal proofs.	X	X		X	X	X	X	X	X	X	X	X	X	X	X	X	X	X	X	X

Standard 8: Number operations	1	2	3	4	5	6	7	8	9	10	11	12	13	14	15	16	17	18	19	20
Mental computation.		X	X	X	X		X	X	X	X	X	X	X	X	X	X	X	X	X	
Standard arithmetic operations.	X	X	X	X	X		X	X	X	X	X	X	X	X	X	X	X	X	X	X
Computational strategies.	X	X	X	X	X		X	X	X	X	X	X	X	X	X	X	X	X	X	X

Standard 11: Patterns, relationships, and functions	1	2	3	4	5	6	7	8	9	10	11	12	13	14	15	16	17	18	19	20
Explaining and analyzing patterns.	X	X	X	X	X	X	X	X	X	X	X	X	X	X	X	X	X	X	X	
Discovering rules.	X	X	X	X		X	X	X	X	X	X		X	X	X	X	X	X	X	X
Relations with two variables.	X	X	X	X	X	X	X	X	X	X	X	X	X	X	X	X	X	X	X	X
Using tables, rules, and equations.	X	X	X	X	X			X			X	X	X			X	X	X		
Problem solving.	X	X	X	X	X	X	X	X	X	X	X	X	X	X	X	X	X	X	X	X
Relationships of variables.	X	X	X	X	X	X	X	X	X	X	X	X	X	X	X	X	X	X	X	X

Standard 13: Algebra	1	2	3	4	5	6	7	8	9	10	11	12	13	14	15	16	17	18	19	20
Standard algebraic notation.	X	X	X	X	X	X	X	X	X	X	X	X	X	X	X	X	X	X		X
Number patterns.	X	X	X	X	X			X	X	X	X				X	X				X
Using variables.	X	X	X	X	X	X	X	X	X	X	X	X	X	X	X	X	X	X	X	X
Simplifying algebraic expressions.	X	X	X		X	X	X	X	X	X	X	X	X	X	X	X	X	X		
Solving equations.										X		X					X			
Operations and properties of numbers.	X	X	X	X	X	X	X	X	X	X	X	X	X	X	X	X		X	X	
Using inequalities.					X		X				X	X						X		

All of the tricks and the accompanying materials have been developed to help support Learning Environment Standard 17:

1. Demonstrate confidence as mathematical thinkers, believing that they can learn mathematics and can achieve high standards in mathematics, and accepting responsibility for their own learning of mathematics.
2. Recognize the power that comes from understanding and doing mathematics.
3. Develop and maintain a positive disposition to mathematics and to mathematical activity.
4. Participate actively in mathematical activity and discussion, freely exchanging ideas and problem-solving strategies with their classmates and teachers, and taking intellectual risks and defending positions without fear of being incorrect.
5. Work cooperatively with other students on mathematical activities, actively sharing, listening, and reflecting during group discussions, and giving and receiving constructive criticism.
6. Make conjectures, pose their own problems, and devise their own approaches to problem solving.
7. Assess their work to determine the effectiveness of their strategies, make decisions about alternate strategies to pursue, and persevere in developing and applying strategies for solving a problem in situations where the method and path to the solution are not at first apparent.
8. Assess their work to determine the correctness of their results, based on their own reasoning, rather than relying solely on external authorities.

THE MISSING EIGHT

Overview of the Trick

The Trick

Write the number 12345679 on a chalk- or whiteboard. A volunteer comes to the board and puts an x over the number he/she likes least. Ask the volunteer to multiply the original number 12345679 by a specific number. The result is a nine-digit number all of whose digits are the number that the volunteer did not like. Yuck.

Example

Suppose the volunteer indicates a dislike for the number 7 by putting the x over the 7. When you ask them to multiply the original number by $9 \times 7 = 63$ they find that:

$$\begin{array}{r} 12345679 \\ \underline{\times 63} \\ 37037037 \\ \underline{+\ 740740740} \\ 777777777 \end{array}$$

How it Works

All conjurers know special properties of the number 12345679. This trick utilizes the fact that $12345679 \times 9 = 111111111$ to make the disliked number appear repeatedly.

Main Concept

Associative property of multiplication

Required Skills

Basic computation	Expanded notation
Multiples	Associative property
Use of variables	Use of algebraic expressions

Teaching and Performing the Trick

Suggested Modes of Instruction

Direct teaching, whole class, or small groups

Materials Required

Chalk/Whiteboard or overhead projector with markers

Performing the Trick – Small groups illustration

1. Divide the audience into small groups. Choose one volunteer from each group and briefly accompany this group of apprentice mathemagicians to an isolated location away from the audience. Explain to them that they will help conjure up a multiplication problem whose result is their groups' least favorite single-digit number. All they have to do is supply one factor in a multiplication problem you will give. They determine this factor by mentally multiplying the least favorite single-digit number that the group chose by 9. Be sure they are comfortable doing this multiplication mentally. Return the apprentices to their groups.
2. Write the number 12345679 on the chalkboard. Note the absence of the digit 8. Have each group write down this number on a piece of paper.
3. Ask each group to determine which digit in this number they like least. Have them put an "x" above the digit that they like the least.
4. Announce that the apprentice mathemagician in each group will supply another number.
5. Ask each group to multiply the two numbers, 12345679 and the number they were given by the apprentice mathemagician in their group. The product will be a nine-digit number all of whose digits are the disliked number.
6. You can repeat the trick or simply have students begin comparing results. Students should see a pattern to the multipliers fairly quickly.
7. As the groups begin to see this pattern, you can suggest that the trick be checked for each possible choice of offending digits. When they have completed all these examples, the class will have given a proof by exhaustion.

THE MISSING EIGHT

8. Hand out the *Student Activity Page* and have the groups work through the algebraic analysis cooperatively.

Analysis – Using algebra to explain why it works

The small groups illustration above demonstrates a proof by exhaustion where all of the possible outcomes are checked. In this case there are only eight possibilities, so a proof by exhaustion is relatively straightforward. When the number of cases becomes too large such a proof can become exhausting unless one enlists the help of a computer and/or calculator.

There is also an algebraic proof that is as follows. Note that $12345679 \times 9 = 111111111$. Now, suppose n represents the digit marked with an "x" in the trick. Verification of the trick is simply based on the multiplication fact above and the associative property of multiplication. When you suggest multiplying by $9n$, the following arithmetic results:

$12345679 \times 9n = (12345679 \times 9) \times n$
$= (111111111) \times n = nnnnnnnnn$

Further Investigations

Students investigate the role of the number 12345679 and discover related patterns which are based on special properties of the number 9.

THE MISSING EIGHT

Name:

Steps in the Trick

1. Write down the number 12345679.
2. Ask a volunteer to put an "x" above the digit that they like the least.
3. Ask the volunteer to multiply 12345679 by a number that you select.
4. The product will be a nine-digit number all of whose digits are the disliked number.

Unraveling the Trick

1. Multiply **12345679** by **9** and observe the result.
2. Choose a variable to represent the number marked with an "x."
3. Write the algebraic expression for **12345679** multiplied by **9** times the number marked with an "x." Use this result to explain the trick.

Extending Questions

1. What numerical values could be used for your chosen variables? What values could not be used?
2. Try the trick using the number **123456789**. Can you predict what digit was marked with the "x"? Explain why or why not.
3. Modify the trick by having the subject subtract the disliked digit from **9**, multiply this difference by **9** and multiply this result by **12345679**. If the final result is announced to you, could you predict which digit was marked with an "x"? How?

Further Investigations

1. In **The Missing Eight** trick, determine what happens to the products if **12345679** are replaced by **123456789**.
2. Develop a way to use a standard eight-place display calculator to compute the products **12345679 × 63** without causing a calculator overflow.
3. Show that the following rule works for squaring any number made up entirely of **9**s;

   For the answer, put down a **1** for the unit's digit; to its left put zeros (one less than the number of **9**s in the number you are squaring); next put an **8**; and finally, **9**s (one less than the number of **9**s in the number you are squaring). Thus, $99^2 = 9801$.
4. Multiply **9** by each number in the following sequence: **21, 321, 4321, 54321, 654321, 7654321, 87654321, 987654321**. Continue until you discover a pattern in the products.

THE HUMAN CALCULATOR

Overview of the Trick

The Trick

Ask a volunteer to write down a five-digit number. Write a number beneath his/her number. This continues until there are several pairs of numbers, with the volunteer announcing when they have written their last number. As soon as they announce that they have written their last number, you write your last number and then, like lightning, you magically write down the correct sum of all of the numbers.

Example

Suppose eight numbers are chosen and the volunteer chooses, at the appropriate time, the numbers 65421, 52578, 97351, and 23451. Your first three choices are 34578, 47421, and 2648. Your fourth choice is somewhat arbitrary, say 31243. The sum you have to compute, quick as lightning, is

65421 + 34578 + 52578 + 47421 + 97351 + 2648 + 23451 + 31243

But this sum is simply:

99999 + 99999 + 99999 + 23451 + 31243

= 100000 + 100000 + 100000 + 23451 + 31243 – 3 = 354691

where the digits of your final choice have been chosen so there was no carrying in the sum of the last two numbers in the list and the unit's digit chosen to cancel out the three that needed to be subtracted.

How it Works

There are many variations of this trick where the mathematics or the performance can be slightly varied to create a new trick. All are based on the magician choosing numbers that complement those chosen by the volunteer so the sums can be computed with ease. In the example above, the magician chooses the first three numbers so that the first three pairs each add to 99999 giving a sum 300000 – 3. The complete sum is then easily determined once the last two numbers are added.

Main Concept

Associative property of addition

Required Skills

Basic computation
Associative property of arithmetic
Use of variables
Use of algebraic expression
Polynomials
Simplifying polynomials

Teaching and Performing the Trick

Suggested Modes of Instruction

Small groups and direct teaching

Materials Required

Chalk/Whiteboard or overhead projector with markers are required. Calculators to check the results of computations are optional.

Performing the Trick – Direct teaching illustration

1. Have a volunteer come to the front of the room and write down a five-digit number.
2. Quickly write down a number under the volunteer's number. Choose your number so that when each corresponding digit of your number and the volunteer's number are summed the result is 9.
3. Repeat Step 2 twice more.
4. Ask the volunteer to write one more five-digit number.
5. Quickly write down a number, choosing it so there is no carrying when added to the volunteer's final number, and so the final digit is a 3.
6. Immediately announce the sum of all of the numbers. This sum will be a six-digit number whose digits are as follows:

    a. Hundred thousand's digit: 3.

THE HUMAN CALCULATOR

b. Ten thousand's digit: sum of the ten thousand's digits in final two numbers chosen.

c. Thousand's digit: sum of the thousand's digits in final two numbers chosen.

d. Hundred's digit: sum of the hundred's digits in final two numbers chosen.

e. Ten's digit: sum of the ten's digits in final two numbers chosen.

f. One's digit: the one's digit of the final number chosen by the volunteer.

7. Repeat the trick with a different volunteer.

8. Distribute the *Student Activity Page* and work through it with the students.

Analysis – Using algebra to explain why it works

Suppose the subject chooses to write a total of $n + 1$ ($n = 3$ in the example above) five-digit number. You follow each of the first n numbers with a number that is 99999 minus the subject's choice. This difference is easily determined digit by digit since there is no carrying. The sum of these $2n$ numbers is $n(99999) = n(100000) - n$. Suppose the last two numbers written are y and z. The sum of the complete column of $2n + 2$ numbers will be:

$$n(100000) - n + (y + z) = n(100000) + (y + z) - n$$

Thus, you add the last two numbers $(y + z)$, subtract the number of pairs above the last two rows (n), and then affix the digit n in front of this result.

Further Investigations

Students investigate generalizations and limitations of this trick before considering links to the arithmetical complement and its potential in subtraction algorithms.

THE HUMAN CALCULATOR

Name:

Steps in the Trick

1. Ask a volunteer to write down a five-digit number.
2. You write down a number under his/her number.
3. Alternate writing numbers like this until there are at least 8 numbers.
4. As soon as the final number is written, ask the volunteer to find the sum of this column of numbers. Quick as lightning, while the volunteer has just begun the computation, you provide the answer.

Unraveling the Trick

1. Assume the subject selects ***n*** five-digit numbers and you follow each selection with a number that will make the sum of the two equal **99999**. Write an algebraic expression for the sum of the ***n*** numbers.
2. Now replace **99999** with **(100000 – 1)** and use the distributive property to remove the parentheses.
3. Choose two different variables to represent the last two numbers selected by the subject and by you.
4. Compute the sum of the expressions in **2** and **3** above. Use this sum to explain the trick.

Extending Questions

1. Can you change the number of five-digit numbers that are selected by the subject? How would this change affect your prediction?
2. Are there any restrictions on the last two numbers selected by the subject? Are there any restrictions on your last choice?
3. What would happen if the subject selected **99999** as the first five-digit number? For all your selections to be five-digit numbers, what restrictions must be placed on the subject's selections?
4. How would your prediction be altered if four-digit numbers were selected instead of five-digit numbers?
5. How could you predict the final sum if you omitted selecting your last number?

Further Investigations

1. Alter the trick so that three-digit numbers are used and so that you will know the sum of the numbers before the subject selects his first number.
2. In the original version of the trick, if someone chose **99999** as a number, this would give away the trick. How can you alter the trick to avoid this situation?
3. In nineteenth century arithmetic texts, the arithmetical complement of a number is defined as the remainder found by subtracting the number from the next highest multiple of **10**. Thus, the arithmetical complement of **8548** is **1452**, since **8548 + 1452 = 10000**. Alter ***The Human Calculator*** trick by incorporating the arithmetical complements of the subject's numbers.
4. Some early arithmetic texts suggested that the arithmetical complement be used in place of subtraction. To subtract two numbers, add the arithmetical complement of the subtrahend to the minuend and then reduce this sum by the multiple of **10** used in obtaining the arithmetical complement.

   For example,

$$\begin{array}{r} 756 \\ -\underline{321} \end{array} \quad \text{becomes} \quad \begin{array}{r} 756 \\ +\underline{679} \\ 1435 \\ -\underline{1000} \\ 435 \end{array}$$

   Compute these differences using addition of arithmetical complements:

$$\begin{array}{r} 546 \\ -\underline{324} \end{array} \qquad \begin{array}{r} 3947 \\ -\underline{1768} \end{array} \qquad \begin{array}{r} 5721 \\ -\underline{846} \end{array}$$

FORTELLING A SUM

Overview of the Trick

The Trick

Ask a volunteer to write down two whole numbers in a column. The volunteer finds the sum to give a third number. He/she continues to add the last two numbers in the column to form another until reaching the seventh number. After a pause to make sure everybody in the class understands the procedure, ask the volunteer to find three more numbers in this column. As they set to work, interrupt to announce that when completed the sum of the ten numbers will be ______. Upon completion of the column, students compute the sum and find, presto!, that you correctly foretold the sum.

Example

Suppose the first two numbers selected are 25 and 17. The ten computed numbers and their sum are then:

1.	25
2.	17
3.	42
4.	59
5.	101
6.	160
7.	261
8.	421
9.	682
10.	+ 1103
The Sum =	2871

But you already knew the sum after the seventh step by mentally computing the product

$$11 \times 261 = 2871$$

How it Works

The sequence of numbers that is created is a *Fibonacci* sequence. Such sequences have many amazing properties. The one that is utilized here is attributed to the magician *Royal Vale Heath*.

Main Concept

Simplifying polynomials

Required Skills

Basic computation
Use of variables
Algebraic expressions
Distributive property
Polynomial addition and subtraction
Simplifying polynomials

Teaching and Performing the Trick

Suggested Modes of Instruction

Small groups, discovery learning, or direct teaching

Materials Required

Chalk/Whiteboard or overhead projector with markers

Performing the Trick – Direct teaching illustration

1. Choose a volunteer and have him/her write down any two whole numbers in a column on a chalkboard, whiteboard, or overhead projector.
2. Have another volunteer write the sum of the two numbers underneath them.
3. Have this volunteer sum the second and third numbers, writing them as the fourth number in the column.
4. Instruct the volunteer to continue until the column contains seven numbers. Encourage the other students to check the arithmetic.
5. Instruct the volunteer to continue this pattern until the column contains ten numbers.
6. As soon as you have given this instruction, mentally calculate the product of the seventh number with 11. Immediately announce, "By the way, when you get all ten numbers listed their sum will be ______."

FORETELLING A SUM

7. Have the volunteer add up the ten numbers, with other students checking the arithmetic. Your prediction will be correct.
8. Hand out the *Student Activity Page* and work through it with students.

Analysis – Using algebra to explain why it works

Let x be the first number chosen and y the second. The list of the ten numbers with their sum is shown below:

$$
\begin{array}{lrcl}
1. & x & = & 1x \\
2. & y & = & 1y \\
3. & x + y & = & 1x + 1y \\
4. & [y + (x + y)] & = & 1x + 2y \\
5. & [(x + y) + (x + 2y)] & = & 2x + 3y \\
6. & [(x + 2y) + (2x + 3y)] & = & 3x + 5y \\
7. & [(2x + 3y) + (3x + 5y)] & = & 5x + 8y \\
8. & [(3x + 5y) + (5x + 8y)] & = & 8x + 13y \\
9. & [(5x + 8y) + (8x + 13y)] & = & 13x + 21y \\
10. & [(8x + 13y) + (13x + 21y)] & = & 21x + 34y \\
 & \text{Sum} & = & 55x + 88y
\end{array}
$$

Notice that the sum, $55x + 88y$, is equal to $[11 \times (5x + 8y)]$ which is just 11 times the seventh term.

Computing a factor of 11 mentally is fairly easy. For example, using the seventh sum in the example above, we need to compute 11×261. But $11 \times 261 = (1 + 10) \times 261 = 261 + 2610 = 2871$. This works for any multiple, the only difficulty can be when carrying is required.

Further Investigations

Students may investigate whether this trick can be performed with numbers other than positive integers. The trick provides a perfect opportunity to have students discover the surprising, ubiquitous Fibonacci numbers.

FORETELLING A SUM

Name:

Steps in the Trick

1. Write down any two numbers in column form.
2. Find the sum of these two numbers and write it in the column below the other two.
3. Find the sum of the second and third numbers and write the sum as the fourth number in the column.
4. Continue adding the last two numbers to form another entry in the column until you have a column of ten numbers.
5. The sum of the ten numbers can be predicted from the seventh number.

Unraveling the Trick

1. Choose two different variables to represent the first and second numbers chosen.
2. Write the algebraic expressions for the other eight numbers in the series using the variables chosen.
3. Sum the ten algebraic expressions.
4. Explain the trick by comparing the computed sum with the seventh algebraic expression.

Extending Questions

1. Does it matter how many digits there are in the first two numbers chosen?
2. What would happen if the two numbers chosen were both zero? Could one of them be zero?
3. What positive integers evenly divide the sum of the ten algebraic expressions?
4. What would be the result if the same variable was used for the two chosen numbers?
5. Can you devise a method to multiply **11** times a number so that the prediction step in the trick becomes a quick, mental calculation?

Further Investigations

1. In the algebraic analysis, do x and y have to be whole numbers for this trick to work? Will it work for integer values of x and y? Rational values?
2. In the algebraic solution you derived under Unraveling the Trick, the coefficients of y are: **1**, **1**, **2**, **3**, **5**, **8**, **13**, **21**, **34**. (Notice that each number after the first two is the sum of the two preceding numbers.) This is the beginning of the series which was published by (and named for) the Italian mathematician *Leonardo Fibonacci* in 1202. List **20** terms of the *Fibonacci* series in a column. Have someone draw a horizontal line under a number in the series and sum all the numbers above the line. You will immediately be able to predict the sum by making a minor alteration to the number in the series which is two steps below the horizontal line. Try it. What is your prediction? Does it always work?
3. Try the prediction trick you just discovered in Investigation #2 above on the series that you used to do the trick. How could your prediction be modified so that the numbers two steps below the horizontal line allows you to predict the sum above the horizontal line? Can this prediction be generalized so that you can predict the sum for any series where the subject selects the first two numbers? Can you explain algebraically how this prediction works?
4. Locate a reference to *Fibonacci* in a mathematics textbook or reference book and explore other properties of this famous series. Additionally, the *Fibonacci* numbers occur frequently in nature. Explore several such instances.

UNVEILING EVEN AND ODD

Overview of the Trick

The Trick

Place one nickel and one dime on a desk. Ask two volunteers to secretly select one coin each from these two coins and hide them in their fist while another volunteer ensures that you are not observing the volunteers' choices. Instruct the first volunteer to double his/her denomination, the second volunteer to triple his/her denomination, and then, continuing in secret, to add the result of these computations together. Then ask them to report whether their result is even or odd. Proclaim which student has selected the nickel and which the dime, confidently inviting the volunteers to certify your magical prowess by unveiling their choices.

Example

Suppose volunteer A selects the dime and volunteer B the nickel. A computes $2 \times 10 = 20$ and B, $3 \times 5 = 15$. They report that the sum of these products $(20 + 15 = 35)$ is odd. From this information you will know that B (who multiplied by 3) selected the odd denomination, 5.

On the other hand, if A selected the nickel and B the dime then the sum of the products, $(2 \times 5) + (3 \times 10) = 40$, would be even. From this information you will know that A (who multiplied by 2) selected the odd denomination, 5.

How it Works

This trick is presented exactly, albeit in more modern language, as it appears in an arithmetic text published in 1760.

(Wingate, Edmund. A treatise of common arithmetic. London: S. Crowder and Company, 1760. p. 370)

The trick uses number theoretic properties of even and odd numbers to secretly encode the volunteer's choices in the result of their computation.

Main Concept

Algebraic expressions

Required Skills

Mental computation
Even and odd numbers
Multiples
Prime numbers
Use of variables
Algebraic expressions
Polynomials

Teaching and Performing the Trick

Suggested Modes of Instruction

Direct teaching, small groups, or whole class

Materials Required

Nickels and dimes

Performing the Trick – Whole class illustration

1. Have the audience divide up into pairs. Give each pair one nickel and one dime. Turn your back and ask each member of each pair to secretly choose one of the two coins and hold them in his or her hand.
2. Turn back around. Tell the group that the person on the left is to multiply their denomination by 2 and the person on the right is to multiply their denomination by 3.
3. Ask the pairs to secretly add these products together.
4. Move from pair to pair asking whether the sum of their products is even or odd. If the sum is odd then the person on the right chose the nickel. If, on the other hand, the sum is even then the person on the left chose the nickel.
5. As you continue to move around, students should begin to see a pattern. As they do, invite them to make predictions.
6. Distribute the *Student Activity Page* and work through it with the students.

UNVEILING EVEN AND ODD

Analysis – Using algebra to explain why it works

- Verification of this trick is based upon the following properties of whole numbers:

An even number plus an odd number is odd; An even number plus an even number is even; An even number times any number is even; An odd number times an odd number is odd.

- Let x and y be the two numbers to be chosen, where one is even and the other odd. Suppose the first person chooses x. Since the first person (A in the example above) multiplies his/her number by 2, this product, $2x$, is always even. When the second person (B in the example above) multiples his/her number by 3, the product, $3y$, will be even if y is even and will be odd if y is odd. If you know that the sum of their two products, $2x + 3y$, is even then you conclude that y must have been even. This tells you that the second person must have chosen the even number. If, however, the sum $2x + 3y$ is odd, then you can conclude that y must have been odd. This tells you that the second person must have chosen the odd number.

This trick can also be verified by examining two separate cases:

Case (1)
The first person selects the even number, the second person the odd number, and:

Case (2)
The first person selects the odd number and the second person the even number.

	$2x$	$3y$	$2x + 3y$
x = even y = odd	even	odd	odd
x = odd y = even	even	even	even

Further Investigations

Students investigate variants and generalizations of this trick.

UNVEILING EVEN AND ODD

Name:

Steps in the Trick

1. **The first person chooses either the nickel or the dime and multiplies its value by 2.**
2. **The second person takes the other coin and multiplies its value by 3.**
3. **They compute the sum of the two products.**
4. **From this sum you can determine which person chose the nickel and which chose the dime.**

Unraveling the Trick

1. Choose two variables to represent the odd and even numbers to be selected.
2. Write algebraic expressions for the products formed by the first and second person and then add these two expressions.
3. If the number selected by the second person was even, will the sum in **2** be even or odd?
4. If the number selected by the second person was odd, will the sum be even or odd?
5. Use your observations in **3** and **4** to explain the trick.

Extending Questions

1. Try replacing the multipliers **2** and **3** by **4** and **5**. Will the trick still work?
2. Could you use any pair of even and odd numbers for multipliers?
3. Would the trick work if addition was replaced by subtraction? Addition by multiplication?
4. Try using negative integers for the even and odd numbers selected by the subjects. Does the trick still work?

Further Investigations

1. Verify the trick using the "two case method" (i.e., in one case the first person chooses the odd number, in the next case, the even number).
2. The 1760 version of this trick refers to devising a "more secret" way for discovering whether the sum is even or odd. Devise such a way.
3. Will this trick work if the two persons are allowed to decide upon their own even and odd numbers?
4. Verify the following version of this trick.

   Someone holds two slips of paper, a slip with an odd number in one hand and a slip with an even number in the other. Have him multiply the number in the right hand by an odd number, and the number in the left hand by an even number. Have him tell you the sum of the two results. If the sum is even, the even number is in the right hand; if the sum is odd, the even number is in the left hand.
5. Instead of using a nickel and a dime for this trick, can you devise a way to use two coins of different denominations for this trick?
6. Give several numerical examples to illustrate how the following variation of the trick works. Note that this variation is not based on whether the two numbers are even or odd.

   Suppose that the two numbers chosen are ***x*** *and* ***y****. Also, suppose* ***x*** *is divisible by a known prime,* ***p****, and* ***y*** *has no common factors with the number* ***x****. After the two people,* **A** *and* **B***, select the numbers* ***x*** *and* ***y****, have* **A** *multiply his/her number by a prime,* ***q****, which is different from the prime* ***p****. Then have* **B** *multiply by the prime* ***p****. Have them tell you the sum of these two products. If the sum is divisible by* ***p****, then* **A** *chose the number* ***x****. If the sum is not divisible by* ***p****, then* **B** *chose* ***x****.*

COUNTERMOVES

Overview of the Trick

The Trick

Ask a volunteer to form three equal Piles of counters, out of your view, from a large collection and place the remainder in a storage container. Ask the volunteer to choose a small number, call it n, which they announce. In sequence, ask the volunteer to move counters as follows: move n counters from the left Pile to the middle Pile; move n counters from the right Pile to the middle Pile; remove the left Pile, placing its counters in the storage container; remove from the middle Pile as many counters as there are in the right Pile, placing them in the storage container; remove the right Pile, placing its counters in the storage container. You then miraculously announce the number of counters that remain.

Example

Suppose that you place 50 counters on an overhead. While your back is turned to the overhead, suppose that the volunteer chooses to make three Piles of 12 counters, returning the 14 remaining counters to the storage container. When you ask for a small positive integer, suppose the volunteer chooses $n = 4$. Then the steps from the trick proceed as follow:

left Pile	middle Pile	right Pile
12	12	12
$12 - 4 = 8$	$12 + 4 = 16$	12
8	$16 + 4 = 20$	$12 - 4 = 8$
$8 - 8 = 0$	20	8
0	$20 - 8 = 12$	8
0	12	0

When the steps are complete, you announce that the number of counters left on the table is $3 \times n$, in this case, $3 \times 4 = 12$.

How it Works

Because the volunteer is allowed to freely choose both the number of counters in the Piles and the small number to be shifted, it seems they have control over the moves. However, the directed moves script the steps to an algebraic identity whose result is always three times the shifting number.

Main Concept

Polynomial addition and subtraction

Required Skills

Basic computation
Multiples
Use of variables
Algebraic expressions
Polynomials
Simplifying polynomials
Polynomial addition and subtraction
Distributive property in algebra

Teaching and Performing the Trick

Suggested Modes of Instruction

Direct teaching, whole class, small groups, or cooperative groups

Materials Required

One or more containers of at least 40 counters or other small, uniform objects

Performing the Trick – Cooperative groups illustration

1. Divide the audience into several small groups. Give each group a container of counters. Have each group shield their working space (desk, table, or bookshelf) from your view.
2. Ask each group to form three equal Piles of counters, removing any unused counters to the container.
3. Have each group choose a small positive integer n (smaller than the number of counters in each Pile) and write this number on a scrap of paper which is also out of your view.
4. Out of your view, have each group make the following counter moves:

COUNTERMOVES

a. Move n counters from the left Pile to the middle Pile. (Remember, n is the small number chosen.)

b. Similarly, move n counters from the right Pile to the middle Pile.

c. Remove the left Pile, placing its contents into the container.

d. From the middle Pile, remove into the container as many counters as there are in the right Pile.

e. Remove the right Pile, placing its contents into the container.

5. In turn, ask each group to report their number n. As they do, divine that they have $3n$ counters left in their working space.

6. Students will begin to see the pattern. As they consider the mechanism for this result, hand out the *Student Activity Page* and have them work through it cooperatively in groups.

Analysis – Using algebra to explain why it works

Suppose that the volunteer places x counters in each of the three Piles and suggests n as the small positive integer in Step 3 of the trick. It is required that $x > n$.

Then results of each of the stages in Step 4 of the trick are as follows:

stage	left Pile	middle Pile	right Pile
3	x	x	x
4a	$x - n$	$x + n$	x
4b	$n - n$	$x + 2n$	$x - n$
4c	0	$x + 2n$	$x - n$
4d	0	$\text{x} + 2n - (x - n)$	$x - n$
4e	0	$\text{x} + 2n - (x - n)$	0

Note that the numbers of counters left on the table are the number in the middle Pile:

$$x + 2n - (x - n) = 3n$$

This is the number you announce to each group.

Further Investigations

Students investigate ways this trick can be adapted to form other tricks based on the same underlying mathematical principles.

COUNTERMOVES

Name:

Steps in the Trick

1. Give a volunteer a large number of counters or other small, uniform objects.
2. Ask the volunteer, out of your view, to form three equal Piles of counters, removing any unused counters to the container.
3. Ask the volunteer to choose a small positive integer n (smaller than the number of counters in each Pile), revealing this number to you.
4. Out of your view, have the volunteer make the following counter moves:
   a. Move n counters from the left Pile to the middle Pile. (Remember, n is the small number chosen.)
   b. Similarly, move n counters from the right Pile to the middle Pile.
   c. Remove the left Pile, placing its contents into the container.
   d From the middle Pile, remove into the container as many counters as there are in the right Pile.
   e. Remove the right Pile, placing its contents into the container.
5. At this point you can predict how many counters remain based on the volunteer's choice of the number n.

Unraveling the Trick

1. Choose one variable to represent the number of counters in each Pile.
2. Use the variable in **1** and the variable n from above to make a table which will indicate the number of counters left in each Pile after each stage of the trick. In the first row, indicate how many counters there are in the respective Piles as **Step 4** of the trick begins.
3. Complete the table in **2** by writing expressions which represent the number of counters in each of the Piles after each of the stages a–e in **Step 4** of the trick.
4. Use the last row of your completed table to explain the trick. What should your final prediction be for the number of counters that remain?

Extending Questions

1. What would happen in the trick if the value of the variable n was greater than the value of the variable?
2. Try the trick when the values of the two variables are equal. Will your prediction still be valid?
3. Could you change the order (or alter) stages a–e in the trick?

Further Investigations

1. Devise a ***Countermove*** trick that starts with a row of five Piles, each containing the same number of counters.
2. Analyze the following trick:

   *Given **9** counters on a table, place **1** in your right hand (call this Pile 1) and one in your left hand (call this Pile 2). Now, starting with your right hand (Pile 1), alternate hands in picking up the counters one at a time until there are none left on the table. With the counters concealed in your hands, begin placing the counters back on the table one at a time. When returning the counters, begin with the left hand and alternate hands until **7** counters have been returned to the table. How many are left in each Pile (hand)?*
3. Given **15** red counters and **15** green counters, place them in a line so that when every ninth counter is discarded, the only counters remaining on the table are the **15** red ones.

PREDICTING THE HOUR

Overview of the Trick

The Trick

Ask a volunteer to think of any number on the dial of a clock and then to add one to the number he/she selected each time you point to a number on the clock with your wand, announcing when he/she reaches 20. When the volunteer announces 20, your pointer will magically be on the number selected by the volunteer.

Example

Suppose the volunteer thinks of six o'clock. As you point to the sequence of numbers shown below, the subject counts silently from the selected number, 6.

Your pointer	Silent count		Your pointer	Silent count
3	7	*Eighth number* ➡	12	14
9	8		11	15
4	9		10	16
8	10		9	17
11	11		8	18
4	12	*Selected number* ➡	7	19
2	13		6	20

You are pointing to 6 when the subject announces the count of 20.

How it Works

As long as you point to the number 12 on the eighth number and work backwards from there, you will be pointing to the hour thought of by the subject. This trick is based on modulo 12 misdirection.

This trick appears in many early mathematics and magic books. It can be used to predict the hour a person was born, when they go to bed, or any other hour suggested by the subject.

Main Concept

Distributive property

Required Skills

Basic computation
Distributive property
Use of variables
Use of algebraic expressions
Polynomials
Simplifying polynomials

Teaching and Performing the Trick

Suggested Modes of Instruction

Small groups or direct teaching.

Materials Required

A 12-hour analog clock and a wand or pointer are required. "Analog clocks" which have a different number of "hours" than 12 are useful for the Further Investigations.

Performing the Trick – Direct teaching illustration

1. Before starting the trick prepare your props by ensuring that your analog clock is easily visible and can be easily reached by you with your wand.
2. Choose a volunteer and have him/her think of any number on the dial of a clock. Have him/her secretly write this number on a scrap of paper and give it to a student to hold.
3. Ask the volunteer to silently add one to the number he/she selected, in succession, each time you point to a number on the clock with your wand, announcing when he/she reaches 20.
4. Slowly point to seven randomly selected numbers on the clock with your wand, providing time for the volunteer to update their count. For the eighth number, point to 12. Then point to 11, 10, 9, . . . in a counterclockwise succession until the volunteer announces that he/she has reached 20.
5. When 20 is announced, your wand will be pointing at the

PREDICTING THE HOUR

chosen number, as can be corroborated by the scrap of paper that was given to the second volunteer.

6. Hand out the *Student Activity Page* and work through it with the students.

Analysis – Using algebra to explain why it works

Suppose the subject is thinking of the number x on the dial. After you point to seven numbers at random, the subject's count is up to $(x + 7)$. Next you point to 12 and the subject's count becomes $(x + 8)$. Since the subject is instructed to count to 20, you will have $20 - (x + 8)$ or $(12 - x)$ more numbers to select before the subject announces that 20 is reached.

You will now move your pointer consecutively backwards from 12, which is equivalent to subtracting the $(12 - x)$ from 12. After $(12 - x)$ times you will be pointing to $12 - (12 - x)$ or x, as the subject announces 20 aloud.

This sequence of computations amounts to the equation $12 - (20 - (x + 8)) = x$, which does not depend upon the subject's choice of x.

Further Investigations

This trick can be generalized to clocks which have a different number of hours than a standard 12-hour clock.

PREDICTING THE HOUR

Name:

Steps in the Trick

1. **Have your subject think of any number on the dial of a clock.**
2. **Ask the person to silently add one to the number he/she selected, in succession, each time you point to a number on the clock, announcing when he/she reaches 20.**
3. **Point to 7 randomly selected numbers on the clock. Point to 12 for the eighth number. Then point to the numbers 11, 10, 9, ... in a counterclockwise succession until 20 is announced.**
4. **When 20 is announced, you will be pointing to the number the subject selected.**

Unraveling the Trick

1. Choose a variable to represent the subject's selected hour.
2. Write algebraic expressions for each of the subject's silent counts as you point to seven randomly selected hours on the clock face.
3. As you point to **12** as your next (eighth) number, write an expression for the subject's silent count.
4. Write an algebraic expression for the difference between the subject's current count and **20**. Simplify this expression.
5. Now you begin moving your pointer consecutively counterclockwise through the counting numbers from **12**. Use the algebraic expression in **4** to explain why you will be pointing to the selected hour when the subject announces **20**.

Extending Questions

1. What are the possible values for the chosen variable? Will the trick work if the subject choose twelve o'clock?
2. If you started by pointing to seven o'clock and moved your pointer consecutively counterclockwise around the clock until the subject announced **20**, would you be pointing to the subject's number?
3. If you were pointing to **12** for the sixth number and the subject counted silently to **18**, would you be able to predict the subject's number?

Further Investigations

1. To vary this trick you can change the count that the subject announces aloud. If you wish to have the subject count to **18** instead of **20**, the trick-equation would become $12 - (18 - (x + 6)) = x$. Thus, you would make sure you are pointing to **12** on the sixth number. When the subject announces that **18** is reached, you would be pointing to x, the number selected by the subject.

   Write the trick-equation if the subject counts to **15**. Explain the procedure for this variation of the game and verify the trick algebraically.
2. You may also vary the game for "clocks" with fewer than **12** hours or more than **12** hours. How would you vary the game for a **10-hour** clock?

AMAZING PROGNOSTICATION

Overview of the Trick

The Trick

A sealed envelope contains a numerical prediction hidden inside of it. Ask a volunteer to announce a random number between 50 and 100. Ask the volunteer to add their number to a specified two-digit number. Then ask them to delete the leftmost digit of this sum and add it to the remaining number. This number is then subtracted from the number originally chosen. *Viola!* The result is the prediction in the sealed envelope.

Example

Suppose your prediction is 43 and your subject chooses 82. Since 99 – 43 = 56, you ask your volunteer to add 56 to their chosen number:

82 + 56 = 138

Deleting the leftmost digit of the sum and adding it to the remaining number yields:

~~1~~38 + 1 = 39

Subtracting this from the original number gives

82 – 39 = 43

How it Works

The key to this trick, a staple of most magicians' repertoires, is that your prediction is safely but secretly encoded in the number you ask the volunteer to add to their number in the first step. When all the computations are completed, your prediction will reappear.

Main Concept

Distributive property

Required Skills

Basic computation
Distributive property
Use of variables
Use of algebraic expressions
Polynomials
Simplifying polynomials
Adding polynomials

Teaching and Performing the Trick

Suggested Modes of Instruction

Direct teaching, whole class, or small groups

Materials Required

Paper, pencils and an envelope for each group are required. Calculators to check computations are optional.

Performing the Trick – Small groups illustration

1. Divide the audience into small groups. Choose one volunteer from each group and briefly accompany this group of apprentice mathemagicians to a location away from the audience. Explain to them that they will correctly foretell the result of their groups' computations. Have each apprentice mathemagician choose a number strictly between 1 and 50. Have them write this number on a piece of paper and place it in a sealed envelope. Have them subtract their number from 99 to form the key to the trick. Return the apprentices to their groups.
2. Have each apprentice mathemagician give their prediction to a member of their group to hold.
3. Ask each group to think of one number strictly between 50 and 100.
4. Ask each group to perform the following calculations:
   a. Add their number to the number the apprentice mathemagician tells them. (This is the key number found in Step 1.)
   b. Delete the leftmost digit from the sum.
   c. Add the deleted digit to the remaining number.
   d. Subtract this result from the number originally chosen.

AMAZING PROGNOSTICATION

5. Have the groups open the envelopes. They will see that the apprentice mathematician's prediction matched the result of their trick in every case.

6. Hand out the *Student Activity Sheets* and have the students work through them in their small groups.

Analysis – Using algebra to explain why it works

This is not much of a trick if you just do it with one volunteer unless you specify the computational aspects in advance. For you clearly know the prediction and if the steps are not specified in advance you can just tell the volunteer to subtract the difference between your numbers from their number. So either specify the computational aspects in advance or have several volunteers.

Let p represent your prediction and n represent a volunteer's choice. Then $1 < p < 50$ and $50 < n < 100$. Your key is $99 - p$. Notice $49 < 99 - p < 98$.

In the computation's first step the volunteer's choice is added to your key, yielding:

$$n + (99 - p)$$

From the conditions set on n and p,

$$99 < (n + (99 - p)) < 198$$

so the result of the first step yields a number with a "1" in the hundred's digit. Deleting this digit then yields:

$$(n + (99 - p)) - 100 = n - p - 1$$

In the third step the deleted digit, 1, is added:

$$n - p - 1 + 1 = n - p$$

In the last step, the volunteer subtracts this number from their original number, resulting in:

$$n - (n - p) = p$$

Further Investigations

Students investigate restrictions and generalizations of this trick.

AMAZING PROGNOSTICATION

Name:

Steps in the Trick

1. Write a prediction (any number strictly between 1 and 50) on a piece of paper and place it in a sealed envelope.
2. Give the paper to someone to hold during the trick.
3. Ask a volunteer to choose any whole number between 50 and 100 and say it aloud.
4. Ask the person who chose the number to perform the calculations below:
   a. Add the two-digit number ____ to the number chosen.
   b. Delete the leftmost digit from the sum.
   c. Add the deleted number to the remaining number.
   d. Subtract this result from the number originally chosen and announce the answer.
5. Ask the person holding the secret paper to read your prediction. It will agree with the result of the computation.

Unraveling the Trick

1. Choose one variable for your prediction and another for the subject's choice.
2. Subtract your prediction from **99** and add this difference to the subject's number (**Step 4a** of the trick).
3. Analyze the algebraic expression in **2** to verify that it represents a three-digit number with a **1** in the hundred's place.
4. Write an algebraic expression after the **1** in the hundred's place has been deleted (i.e. **100** subtracted) and a **1** added to the remaining two-digit number (**Steps 4b** and **4c** of the trick). Simplify this expression.
5. Subtract this simplified expression from the subject's chosen number and use the result to explain the trick.

Extending Questions

1. What are the possible values for your prediction (i.e. the variable p)? What are the possible values for the subject's choice (i.e. the variable n)?
2. Is the deleted digit always a **1**? Explain why by analyzing the expression $99 - p + n$.
3. Try the trick if the subject chooses a number less than **50**. Does it work?
4. Try the trick if your prediction number is greater than **50**. Does it work?

Further Investigations

1. Will this trick work if your prediction is either **1** or **50** and your subject chooses **50** or **100**?
2. Alter this trick so that your prediction is between **100** and **200** and your subject can choose a number between **200** and **1000**.

THE DIVINING ROD

Overview of the Trick

The Trick

Tell the audience that you have a magical "divining rod" which will predict the outcome of a sequence on computations that will be performed on a number chosen by a volunteer. With great fanfare, you provide a volunteer with a sealed envelope containing the secretly transcribed message "ROD = 18 + 15 + 4 = 37." Ask another volunteer to secretly choose a three-digit number with no repeated digits. Write a sequence of computational steps on an overhead: find the sum of the six permutations of the three-digit number, divide the sum by 3, divide the resulting quotient by 2, and, finally, divide the resulting quotient by the sum of the digits of the original three-digit number. Have the volunteer perform these computations–using the audience as helpers if necessary. Once the computation is completed and the result announced, have the first volunteer open the envelope which will correctly have predicted the outcome before the initial number was even chosen.

Example

If the number selected was 457, then the six permutations are as follows:

457 475 547 574 745 754

Their sum is

$457 + 475 + 547 + 574 + 745 + 754 = 3552$

Upon division by 3 and then by 2 the quotients are

$3552 \div 3 = 1184$ and $1184 \div 2 = 592$

Since the sum of the digits of the original number is

$4 + 5 + 7 = 16$

the final division is $592 \div 16 = 37$

and your divining rod has correctly predicted the result of the computations.

How it Works

In earlier times, a divining rod was used to locate underground water. In this trick your divining rod literally predicts ROD. These letters are the 18th, 15th, and 4th letters in the English alphabet respectively and $18 + 15 + 4 = 37$. This will always be the result of the trick, no matter which three-digit number is originally chosen.

Main Concept

Polynomial addition and factoring

Required Skills

Basic computation
Expanded notation of integers
Multiples
Distributive property
Permutations
Divisibility Properties
Use of variables
Use of algebraic expressions
Polynomials
Simplifying polynomials

Teaching and Performing the Trick

Suggested Modes of Instruction

Direct teaching, small groups, or whole class

Materials Required

Paper, pencils and an envelope for each group are required. Calculators to check computations are optional.

Performing the Trick – Whole class information

1. Prior to beginning the trick, secretly write your prediction, "ROD = 18 + 15 + 4 = 37," on a slip of paper and seal it in an envelope.

THE DIVINING ROD

2. Tell the audience that you will ask each of them to choose a three-digit number and perform a sequence of calculations based on these numbers. You have sealed a prediction in a sealed envelope. This prediction is a divining rod that will divine the person who has the predicted answer from the group.
3. Ask each member of the audience to choose a three-digit number with no digits repeating.
4. Ask each member of the audience to follow the steps below, keeping the results of their computations secret:
   a. Make a list of the six different permutations of your three-digit number.
   b. Find the sum of these six permutations.
   c. Divide the resulting sum by 3.
   d. Divide the resulting quotient by 2.
   e. Divide this next quotient by the sum of the digits in the original number.
5. While the audience keeps the result of their computations secret, have the divining rod lead you around the room, theatrically using it as a prop to lead you to an audience member whose answer you will correctly divine. It doesn't matter who you choose to be *lead* to–they all have the same answer. When you arrive at the divined audience member, give them the envelope and have them check to see if it correctly predicted the answer.
6. Explain why "ROD = 37." As you do so, the audience will quickly realize that each of their computations resulting in 37 so there was not much to divine. This should pique their curiosity about how the trick works.
7. Hand out the *Student Activity Page* and work through it with the students.

Analysis – Using algebra to explain why it works

Let the digits in the chosen number be a, b, and c. Note that no two of these variables have the same value. Thus, there are six permutations:

abc
acb
bac
bca
cab
cba

The sums of each of these three columns are:

hundred's column	ten's column	one's column
$2a + 2b + 2c$	$2a + 2b + 2c$	$2a + 2b + 2c$

Thus, the sum, S, expressed in expanded notation is:

$$S = (2a + 2b + 2c)100 + (2a + 2b + 2c)10 + (2a + 2b + 2c)$$
$$= 200(a + b + c) + 20(a + b + c) + 2(a + b + c)$$
$$= 222(a + b + c)$$

The result of the three subsequent divisions yields the following values:

$$S \div 3 = 222(a + b + c) \div 3 = 74(a + b + c),$$
$$74(a + b + c) \div 2 = 37(a + b + c)$$
$$37(a + b + c) \div (a + b + c) = 37$$

Hence, the result is always 37.

Further Investigations

Students determine whether this trick can be verified via proof by exhaustion, investigate possible limitations of this trick and whether it can be generalized to four-digit numbers.

THE DIVINING ROD

Name:

Steps in the Trick

1. **Choose a volunteer and ask him/her to secretly choose a three-digit number with no repeating digits.**
2. **Ask the volunteer to follow the steps below, keeping the results of their computations secret:**
   a. **Make a list of the six different permutations of your three-digit number.**
   b. **Find the sum of these six permutations.**
   c. **Divide the resulting sum by 3.**
   d. **Divide the resulting quotient by 2.**
   e. **Divide this next quotient by the sum of the digits in the original number.**
3. **You can predict the result of their computation without knowing their original number or the result of any of their computations.**

Unraveling the Trick

1. Select three variables to represent the hundred's, ten's, and unit's digits of the chosen number.
2. Using the variables from **1**, write the six possible permutations of the chosen number in a column.
3. Find the sum of the six permutations. (Hint: Separately sum the hundred's, ten's, and unit's digits and then express the sum in expanded notation.)
4. Starting with the algebraic expression for the sum, carry out the three divisions outlined in the trick.
5. Use the results of the last step to explain the trick and determine the prediction.

Extending Questions

1. How many choices are there for the hundred's variable? The ten's variable? The one's variable?
2. Try the trick with a three-digit number with two identical digits. Does the trick work?
3. Try the three divisions in a different order. Does the trick still work?
4. In choosing a three-digit number with three different digits, how many choices does the subject have?

Further Investigations

1. Is it feasible to verify this trick through proof by exhaustion, that is, by listing all possible choices and verifying each separately?
2. Examine what happens if either the ten's digit or unit's digit of the chosen number is zero. Does the trick work?
3. Try modifying the trick so the subject can select a three-digit number where exactly two of the digits are identical. What would happen if three of the digits were identical?
4. Develop a trick that begins with the subject selecting a four-digit number, listing its permutations, and summing them.

PREDICTABLE DICE

Overview of the Trick

The Trick

Give a volunteer two dice. Tell the audience that when the dice are rolled the top and bottom numbers of one of the die are to be multiplied by the top and bottom numbers of the other die. The mystery is to predict the sum of the resulting four products. With great fanfare you provide an eager volunteer with a sealed envelope containing the number 49 written on a sheet of paper. The dice are then rolled, the products and sum computed, and, low and behold, the answer is 49. Just as you predicted!

Example

Suppose a 6 is rolled on Die #1 and a 3 on Die #2. By examining a pair of dice you will see that a 1 appears on the bottom of Die #1 and a 4 appears on the bottom of Die #2.

The four products and their sum are:

$$\begin{aligned} 6 \times 3 &= 18 \\ 1 \times 4 &= 4 \\ 6 \times 4 &= 24 \\ 1 \times 3 &= 3 \\ \text{The sum } & 49 \end{aligned}$$

Presto!

How it Works

No matter what two numbers are rolled, the sum of the four products will be 49.

Main Concepts

Polynomial addition and simplification

Required Skills

Basic computation
Distributive Property
Use of variables
Use of algebraic expressions
Polynomials
Polynomial addition
Simplifying polynomials
Polynomial factoring

Teaching and Performing the Trick

Suggested Modes of Instruction

Direct teaching, whole class, small groups, or cooperative groups

Materials Required

Two regular, six-sided, distinguishable dice for each group of students. An envelope and sheet of paper for each group. Polyhedral dice for the Further Investigations are optional.

Performing the Trick – Cooperative groups illustration:

1. Before the trick starts, secretly write the number 49 clearly on several sheets of paper, placing each inside of a sealed envelope.
2. Divide the class into cooperative groups of three to six students per group.
3. Give each group a pair of six-sided dice.
4. Have each group designate one of the dice as Die #1 and the other as Die #2. For example, white die as Die #1, and red die as Die #2.
5. Have each group roll their dice, keeping the result hidden from your view.
6. Hand each group a sealed envelope, with some dramatic fanfare as if you were trying very hard to conjure the identity of their rolls despite the fact that the contents of all the envelopes are the same.
7. Instruct students that each group is to determine the following products:
   (top of Die #1) × (top of Die #2)
   (bottom of Die #1) × (bottom of Die #2)
   (top of Die #1) × (bottom of Die #2)
   (top of Die #2) × (bottom of Die #1).
8. Instruct students to find the sum of the four products they determined and compare this with the prediction hidden in the envelope. Your prediction will be correct.

PREDICTABLE DICE

9. Hand out the *Student Activity Sheets* and have the students work through them cooperatively in groups.

Analysis – Using algebra to explain why it works

Note that the number which appears on the top of a regular, six-sided die added to the number on the bottom of that die is always 7. Thus, 1 is opposite 6, 2 is opposite 5, and 3 is opposite 4.

When the dice are rolled, let
x = the number on top of Die #1, and
y = the number on top of Die #2.

From the observation above, it follows that the number on the bottom of Die #1 is $(7 - x)$ since $7 + (7 - x) = 7$. Similarly, the number on the bottom of Die #2 is $(7 - y)$. The four products will be:

$$\begin{aligned}(x)(y) &= xy\\ (7-x)(7-y) &= 49 - 7x - 7y + xy\\ x(7-y) &= 7x - xy\\ y(7-x) &= 7y - xy\end{aligned}$$

The sum of these products is:

$$xy + 49 - 7x - 7y + xy + 7x - xy + 7y - xy = 49$$

This algebraic computation shows that this sum does not depend on the values of x and y.

This outcome can also be verified using a proof by exhaustion, since there are a finite number of rolls for a pair of dice (thirty-six in all). One could verify that the sum of the four indicated products is 49 for each of the thirty-six possible rolls.

Further Investigations

Students may investigate ways in which this trick can be extended to polyhedral dice that have become so common in many contemporary children's games.

PREDICTABLE DICE

Name:

Steps in the Trick

1. **Roll a pair of regular, distinguishable, six-sided dice.**
2. **Find the following products: (top of Die #1) × (top of Die #2)**
   (bottom of Die #1) × (bottom of Die #2)
   (top of Die #1) × (bottom of Die #2)
   (top of Die #2) × (bottom of Die #1)
3. **Find the sum of the products.**
4. **The sum of these four products can be predicted before the dice are rolled.**

Unraveling the Trick

1. Assign a variable to the number showing on top of one die and a different variable to the number on top of the other die.
2. Examine a die. What is the sum of the two numbers on opposite sides of the die?
3. Using the variables chosen in **1**, write algebraic expressions for the numbers appearing on the bottom of each die.
4. Find the algebraic expressions for the four indicated products.
5. Find the sum of these four products. Explain how the trick works after algebraically simplifying the sum.

Extending Questions

1. What are the possible values for each of the two chosen variables?
2. If the sum of the numbers on the opposite sides of a die was not **7**, would the trick work? Try it if **1** was opposite **5**, **2** opposite **4**, and **3** opposite **6**.
3. Does your proof work if a double is rolled? (Use ***x*** for the top value of each die.)
4. How many possible outcomes are there when a pair of dice is rolled? Could you verify this trick using a proof by exhaustion?

Further Investigations

1. Polyhedra dice are commercially available. The placement of the numbers on these dice vary, however, depending on the manufacturer. On one pair of dodecahedra dice (**12** pentagonal faces) the numbers **1** to **12** are paired as follows on each die.

   Use the fact that the sum of the top and bottom numbers is always **13** to alter the trick for a pair of dodecahedra dice.

Top	Bottom	Top	Bottom
1	12	4	9
2	11	5	8
3	10	6	7

2. One manufacturer numbers the **20** faces of their icosahedra dice (**20** triangular faces) **0** to **9** twice. The numbers are paired on each die as follows.

Top	Bottom	Top	Bottom
0	9	5	4
1	8	6	3
2	7	7	2
3	6	8	1
4	5	9	0

   Use the fact that the sum of the top and bottom numbers is always **9** to alter the trick for a pair of icosahedra dice.

3. Suppose three standard six-sided dice are rolled and the sum of the following products of two numbers is found:

   (1) The products of all combinations of top numbers appearing on the three dice (i.e., Top 1 × Top 2; Top 2 × Top 3; Top 1 × Top 3);

   (2) The products of all combinations of bottom numbers appearing on the three dice; and

   (3) The products of the top number on each die with each of the two bottom numbers on the other dice.

   Can you predict the sum of these **12** products? Can you prove that your prediction always works?

WHO'S HIDING THE COIN?

Overview of the Trick

The Trick

Assign each member of a group of two to nine people a single-digit identification number. While you are not watching, the group invites one of its members to secretly conceal a coin (or other object) in one of their hands. You then ask a volunteer from the group to mentally perform a sequence of arithmetical computations based on the identifying number of the person holding the coin. Namely: multiply the identifying number by 2; add 5; multiply by 5; add 10; add 1 if the coin is in the right hand, 2 if it is in the left. Asking the volunteer to reveal the result of the computations, you can immediately identify both the person concealing the coin and the hand that is concealing it.

Example

With seven people (A, B, C, D, E, F, and G) you naturally assign the identification numbers 1 to A, 2 to B, 3 to C, etc. Suppose that F has hidden the coin in her left hand. You approach one person at random, say person B, and ask this person to follow the six steps outlined above. It might be necessary to review the identification numbers for the seven people. Person B's computations are then as follows:

$$
\begin{aligned}
2 \times (\text{F's I.D. number}) = \quad 2 \times 6 &= 12 \\
12 + 5 &= 17 \\
5 \times 17 &= 85 \\
85 + 10 &= 95 \\
95 + 2 &= 97
\end{aligned}
$$

When B announces 97 as a result you mentally compute 97 – 35 = 62. The ten's digit, 6, tells you person F has the coin and the ones digit, 2, tells you it is in her left hand.

How it Works

Based on a child's game called *Who wears the ring?*, the volunteer's computation secretly encodes the concealer's identification number in the ten's digit and the hand in the one's digit.

Main Concept

Solving linear equations

Required Skills

Basic mental computation
Distributive law
Use of variables
Use of algebraic expressions
Polynomials
Simplifying polynomials

Teaching and Performing the Trick

Suggested Modes of Instruction

Direct teaching, whole class, small groups, or discovery learning

Materials Required

Coin or other small object that can be easily hidden in the subjects' hands. Several different coins or other distinguishable objects for the Further Investigations are optional.

Performing the Trick – Discovery learning illustration

1. Choose a group of 2 to 9 volunteers to come forward. Have the students accompany you into a location out of sight of the other students. Give one of the volunteers a coin to hold in one of their hands. Return to the room.
2. Have the volunteers count off in order so they all have identification numbers from 1, 2, 3, etc.
3. Tell the class that you are going to do a sequence of computations on this identification number. If they follow along algebraically, calling the identification number x, they should learn who is holding the coin and what hand it is in.
4. Announce the following steps, having the volunteers mentally perform the following computations and giving the other students time to do the algebraic calculations:

WHO'S HIDING THE COIN?

a. Multiply the identification number of the person concealing the coin by 2.

b. Add 5 to this number.

c. Multiply the result by 5.

d. Add 10 to this product.

e. Add 1 to the result if the coin is in the concealer's right hand, 2 otherwise.

5. Have the volunteers announce the final result of their computation.

6. Wait. If members of the audience have completed their algebraic computations correctly, they will be able to solve a simple algebraic equation to divine who has the coin and what hand it is in. The result of the computation is $10x + 35 + (1 \text{ or } 2)$.

7. Hand out the *Student Activity Page* and have the students work through it in groups.

Analysis – Using algebra to explain why it works

Let x represent the identification number of the person concealing the coin (remember, $x = 1, 2, 3, \ldots$) and y represent the hand they are concealing the coin in where $y = 1$ for the right hand and $y = 2$ for the left hand. Then the steps in the computation from stage 4 of the trick are as follows:

Multiply identification number by 2.	$2x$
Add 5 to product.	$2x + 5$
Multiply sum by 5.	$5(2x + 5) = 10x + 25$
Add 10 to product.	$10x + 25 + 10 = 10x + 35$
Add 1 or 2 depending on hand.	$10x + 35 + y$

The number announced will be $10x + 35 + y$. Now, you subtract 35 from the number announced and the answer will be $10x + y$. Note that the unit's digit, y, identifies the hand (1 for right, 2 for left) and the ten's digit is x, the identity number of the person concealing the coin.

The end of this trick requires a slight modification if the number of people exceeds nine.

Further Investigations

This trick can be adapted to work for groups of different sizes and/or where the concealer is allowed to choose from among several different denominations of coins.

WHO'S HIDING THE COIN?

Name:

Steps in the Trick

1. Have members of a group of two to nine people count off so they have identification numbers 1, 2, 3, ...
2. While you are out of view of the group, have one member of the group conceal a coin in one hand. Make sure each member of the group knows where the coin is.
3. Randomly choose a volunteer to mentally perform the following computations, keeping the computations secret as they proceed:
   a. Multiply the identification number of the person concealing the coin by 2.
   b. Add 5 to this number.
   c. Multiply the result by 5.
   d. Add 10 to the result.
   e. Add 1 to the result if the coin is in the concealer's right hand, 2 otherwise.
4. Have the volunteer announce the final result of their computation.
5. From this result, you magically identify the exact location of the concealed coin.

Unraveling the Trick

1. Choose a variable to represent the identification number of the person holding the coin. Choose a second variable to identify which hand holds the coin.
2. Write algebraic expressions for each of the successive computational Steps in the Trick.
3. From the last algebraic expression subtract **35** and use the result to predict who is holding the coin and in which hand the coin is held.

Extending Questions

1. What possible values can be used for each variable?
2. After **35** is subtracted from the announced number, will the result always be a two-digit number?
3. If a number other than **10** was added in the trick's fourth computation, let's say **15**, how would this affect the trick?
4. How would the outcome of the trick be affected if you added **1** for the right hand and did no further computation if the left hand was used?

Further Investigations

1. Try the trick for more than nine people so the identification numbers must include two-digit numbers. Explain the "slight modification" in the trick that is needed to predict the person concealing the coin.
2. Does there seem to be restriction on the number of people that can make up the group concealing the coin?
3. To complicate the trick, allow the person to conceal one of a penny, nickel, dime, quarter, half-dollar, or silver dollar, removing the excess coins from view. Substitute the following steps after **Step3e** in the trick's instructions on the previous page.
   f. Multiply the result by **10**.
   g. Add the appropriate one-digit number.
      1 for a penny
      2 for a nickel
      3 for a dime
      4 for a quarter
      5 for a half-dollar
      6 for a dollar
   h. Announce the result.

   In our example, suppose a quarter was concealed in the left hand of person number **6**. Then **Step f** results in **10 × 97 = 970** and **Step g** gives **970 + 4 = 974**. When this result is announced, you subtract **350** and the answer **624** identifies "person number **6**," "the left hand," and "a quarter is concealed." Use algebra to show why this variation of the trick works.
4. Use algebra to explain how the following version of this trick works. *(Hint: Refer to Trick 4—Unveiling Even and Odd.)*
   a. Ask someone to put a dime in one hand and a penny in the other.
   b. Ask him/her to multiply the value of the coin in his/her right hand by **4**, **6**, or **8**, and the value of the coin in his/her left hand by **3**, **5**, or **7**.
   c. Now ask him/her to add the results and tell you the total. If the total is even, he/she has the penny in the right hand; if the total is odd, he/she has it in the left hand.

SO HOW AM I DOING?

Overview of the Trick

The Trick

Each member of an audience is to sum several numbers that are unique to each individual; e.g. their age on their last birthday, the number of letters in their name, how clever they are on a scale of 1–10, etc. They are then asked to subtract the sum of the digits of this number from itself. They are then asked to sum the digits of the difference. If this sum is not a single digit they continue to sum the digits of the resulting sum until the sum is a single digit. They then square this result, subtract one, and divide by 8. The result is their "final score": how they are doing on a scale of 1–10. In this trick the score is always a 10, but the trick can be readily adapted to yield any value in a given scale.

Example

Suppose one audience member is 14-years-old, has 19 letters in their name, and thinks that they rank as a 9 in terms of how clever they are on a scale of 1–10. The sum of these numbers is 42. Subtracting the sum of the digits yields $42 - 6 = 36$. The sum of the digits of this difference is 9. $9^2 = 81$. Subtract 1 and dividing by 8 gives $(81 - 1) \div 8 = 10$: they're doing great!

How it Works

Subtracting the sum of the digits of any number from this number results in a multiple of 9. The number 9 has many magical properties which we see in many mathematical magic tricks *(e.g. Trick 18–Magical Number Nine)*. Here, since we have a multiple of 9 when we sum the digits we will also have a multiple of 9. Continuing in this way until we reach a single-digit number, the number will be 9. The trick then builds the desired ranking from there.

Main Concepts

Polynomial addition and divisibility properties

Required Skills

Basic computation
Expanded notation for integers
Multiples
Distributive property
Permutations
Use of variables
Use of algebraic expressions
Polynomials
Simplifying polynomials

Teaching and Performing the Trick

Suggested Modes of Instruction

Direct teaching or whole class

Materials Required

Paper and pencils for computations are required. Calculators to check computations are optional.

Performing the Trick – Whole class illustration

1. Ask each member of the audience to write the following numbers in a column:
   a. Their age as of their last birthday.
   b. Their favorite number.
   c. The number of letters in their full name.
   d. How clever they think they are on a scale of 1–10.
   e. Any other numerical quantities unique to the individuals.
2. Ask each audience member to sum the column of figures.
3. Subtract the sum of the digits in this number from the number itself.
4. Compute the sum of the digits of this difference.
5. If the sum of the digits is more than a single digit, find the sum of the digits of this sum. Repeat until the sum of the digits are a single digit.

SO HOW AM I DOING?

6. Square the result.
7. Subtract 1.
8. Divide by 8.
9. This result tells each audience member how well they are doing on a scale such as:

   1. yuck!
   2. not so good
   3. in the middle
   4. kind of good
   5. all right
   6. okay
   7. pretty good
   8. rather good
   9. excellent
   10. great, superb!

Analysis – Using algebra to explain why it works

After Step 2 in the trick, the subject's number, N, will likely be a two- or three-digit number depending on the person's age, etc. Let us suppose it is a three-digit number with digits a, b, and c. Then, we can write N in expanded notation as $N = 100a + 10b + c$. The computation in Step 3 gives:

$N - (a + b + c) = (100a + 10b + c) - (a + b + c)$
$= 99a + 9b = 9(11a + b)$

Thus, the result is a multiple of 9.

If a number is divisible by 9, then so is the sum of its digits. Therefore, the result of Step 4, and Step 5 if necessary, will yield a single-digit multiple of 9. This number must be 9. The last three Steps in the Trick produce the numbers: $9^2 = 81$, $81 - 1 = 80$, and $80 \div 8 = 10$, everyone's score.

Further Investigations

Students investigate ways this trick can be adapted and analyze related tricks which are based on special properties of the number 9.

SO HOW AM I DOING?

Name:

Steps in the Trick

1. Ask a volunteer to write the following numbers in a column:
   a. Their age as of their last birthday.
   b. Their favorite number.
   c. The number of letters in their full name.
   d. How clever they think they are on a scale of 1–10.
   e. Any other numerical quantities unique to the individual.
2. Ask the volunteer to sum the column of figures.
3. Subtract the sum of the digits in this number from the number itself.
4. Compute the sum of the digits of this difference.
5. If the sum of the digits is more than a single digit, find the sum of the digits of this sum. Repeat until the sum of the digits are a single digit.
6. Square the result.
7. Subtract 1.
8. Divide by 8.
9. This result tells the volunteer how well they are doing on a scale such as:
   ❶ *Yuck*
   ❷ *Not so good*
   ❸ *In the middle*
   ❹ *Kind of good*
   ❺ *All right*
   ❻ *Okay*
   ❼ *Pretty good*
   ❽ *Rather good*
   ❾ *Excellent*
   ❿ *Great, superb*

Unraveling the Trick

1. How many digits might there be in the subject's number after **Step 2**?
2. Assume the subject's number, ***N***, has three digits. Choose three variables to represent the three digits in ***N***. Write ***N*** in expanded notation.
3. Write an expression for the difference between ***N*** and the sum of its digits. Simplify and factor the result.
4. What will be the result of **Step 4**, and possibly **Step 5** if necessary, in the trick?
5. Using your observation in **4**, compute the arithmetic steps for the rest of the trick to predict the final score.

Extending Questions

1. Does it make any difference what numbers the subject uses in **Step 1** of the trick?
2. What is special about a number minus the sum of its digits?
3. Alter **Steps 6** to **8** so the final score is still **10**. Does the trick work?
4. Alter **Steps 6** to **8** so the final score is **5**. Does the trick still work?

Further Investigations

1. Alter the Steps in the Trick so that the final answer is a number other than **10**.
2. Verify that if the difference of any three-digit number and the three digits in reverse order is divided by **9**, then the quotient will always be a multiple of **11**.
3. Analyze the following trick:
   a. Write a number with **3** or more digits.
   b. Divide by **9** and tell me the remainder, R_1.
   c. Delete any non-zero digit from the original number.
   d. Divide the resulting number by **9** and tell me the remainder, R_2.

SO HOW AM I DOING?

Name:

e. I can now predict the deleted digit.

How does this trick work?

4. Analyze the following trick
   a. Write down a four-digit number.
   b. Write down the sum of the four digits.
   c. Delete a digit in the original number.
   d. Write the **3** remaining digits above the sum found in **b** above.
   e. Subtract the numbers and announce the result.
   f. Your prediction of the deleted digit is found by adding the digits in the announced number and subtracting this from the nearest multiple of **9**.

Does this always work?

THE MARBLE TRANSFER

Overview of the Trick

The Trick

Outside of your view a volunteer, say a male denoted by A, is asked to take a handful of marbles from a bowl and secretly count them. Another volunteer, say a female denoted by B, is asked to take three times as many marbles from the bowl as A. A is then asked to give 5 of his marbles to B. In return, B is asked to give back to A 3 times as many marbles as A is now holding. You immediately predict that B is holding 20 marbles. After asking how many marbles A currently has you continue your amazing prognostication by announcing how many marbles each of the volunteers took from the bowl.

Example

Suppose A takes 15 marbles. Then B takes 3×15 or 45 marbles. A gives 5 marbles to B, leaving A with 10 and B holding 50. Now B gives $3 \times 10 = 30$ marbles to A, leaving B with 20 and A holding 40. At this point you correctly announce that B has 20 and ask A how many he has. When A says 40, compute $(40 + 20) \div 4 = 15$ and announce that A originally took 15 marbles and B took 45.

How it Works

The second volunteer always ends up with 20 marbles. The number of marbles originally taken by the first volunteer is related to the number they end up with by a simple linear equation that can be easily solved mentally and is, of course, one third of the number first chosen by the second volunteer. This trick is adapted from an early arithmetic recreation.

Main Concept

Solving simple linear equations

Required Skills

Basic computation
Distributive property
Use of variables
Use of algebraic expressions
Polynomials
Polynomial addition
Simplifying polynomials
Solving equations

Teaching and Performing the Trick

Suggested Modes of Instruction

Small groups, direct teaching, or cooperative groups

Materials Required

A bowl or other small container containing at least 80 marbles, or other objects small enough so at least a dozen can be held in a handful, are required.

Performing the Trick – Direct teaching illustration

1. Place the marbles in a small container.
2. Choose two volunteers, which we'll call A and B. Outside of your view ask the volunteers to proceed as follows:
   a. A, take a handful of marbles—at least five—from the container and count them without sharing the result with the magician.
   b. B, take from the container three times as many marbles as A took.
   c. A, give 5 of your marbles to B.
   d. B, give back to A 3 times as many marbles as A is now holding.
3. Dramatically announce that B is holding 20 marbles.
4. While the audience is abuzz, nonchalantly ask A how many marbles he/she has.
5. Add 20 to this number and then divide by 4. This is how many marbles A originally chose. B chose three times this many. You can now announce these initial choices.
6. Divide the class up in small groups and distribute the *Student Activity Page*. Have each group mimic the trick in their group as they are guided to discover how the trick works via tasks on the *Student Activity Page*.

THE MARBLE TRANSFER

Analysis – Using algebra to explain why it works

Let n represent the number of marbles taken by subject A. Subject B takes $3n$ marbles. After the first transfer of 5 marbles, A has $(n - 5)$ marbles and B has $(3n + 5)$ marbles. The second transfer involves $3(n - 5)$ marbles from B to A. After this transfer, A has $(n - 5) + 3(n - 5) = 4n - 20$ marbles and B has $(3n + 5) - 3(n - 5) = 5 + 15 = 20$ marbles. In summary, after the transfers one has:

Number of marbles A and B have at each step.

	A	B
Original draw	n	$3n$
First transfer	$n - 5$	$3n + 5$
Second transfer	$(n - 5) + 3(n - 5) =$ $4n - 20$	$(3n - 5) - 3(n - 5)$ $= 20$

Hence, you are able to announce that B holds 20 marbles.

Let x represent the number of marbles A is currently holding. As soon as you know the value of x, solve the equation: $x = 4n - 20$ for n, the number of marbles originally taken from the bowl. This involves adding 20 to x and dividing the resulting sum by 4. Having correctly found the number of marbles originally taken by A you need only multiply it by 3 to tell B how many she originally took from the bowl.

Further Investigations

The proportion of marbles B chooses as well as the number of marbles that are first transferred from A to B can be varied. Students may investigate a completely general version of this problem.

THE MARBLE TRANSFER

Name:

Steps in the Trick

1. Person A takes a handful of marbles from a container and secretly counts them.
2. Person B takes from the container three times as many marbles as A took.
3. Person A gives 5 of his/her marbles to person B.
4. Person B gives back to A three times as many of his/her marbles as A currently holds.
5. At this point the number of marbles person B is holding can be predicted.
6. When told how many marbles Person A now holds, you can determine how many marbles both A and B originally took.

Unraveling the Trick

1. Choose a variable to represent the number of marbles taken by subject **A** and write an expression for the number of marbles taken by **B** in terms of this variable.
2. Write algebraic expressions for the number of marbles held by subject **A** and by subject **B** at the end of each marble transfer.
3. Simplify the last expressions to explain why subject **B** holds **20** marbles at the end of the trick.
4. Show that you can predict the original number taken by **A** if you add **20** to the final number held by **A** and divide this sum by **4**.

Extending Questions

1. Does the trick depend upon the original number of marbles in the bowl? What would happen if the bowl contained **30** marbles and **A** selected **8**?
2. If there were **30** marbles in the bowl, what is the maximum number of marbles **A** could select and have the trick work?
3. If there were **100** marbles in the bowl to start, what is the maximum number **A** could select for the trick to be completed? What would happen if **A** selected **5** or fewer marbles?
4. Based on Questions 1–3 above, how could the directions be made more specific so that the trick will always work?
5. If there were a fixed number of marbles in the bowl (say **100**), could you verify the trick using a proof by exhaustion?

Further Investigations

1. For what values of n (the number of marbles originally taken by subject **A**) will this trick fail? How can you assure yourself that this will not happen?
2. After the two transfers of marbles take place, you know **B** is holding **20** marbles. Think of another way to end this trick and explain your prediction or predictions.
3. How would your predictions be affected if:
   (a) **B** took **4** (instead of **3**) times as many marbles as **A**; and
   (b) the first transfer of marbles was **7** marbles (instead of **5**); and
   (c) the second transfer was **4** (instead of **3**) times the number now held by **A**?
4. In the original trick, use p as the multiplier (in place of **3**). Also, use q as the first transfer value (in place of **5**). Verify with algebra that the first prediction should be $q\ (1 + p)$. Also, verify that the original number of marbles taken by **A** will be $(x + q(1 + p)) \div (1 + p)$, where x is the number of marbles last held by **A**.
5. Here is another marble transfer trick. Bowl 1 contains **20** white marbles and Bowl 2 contains **20** black marbles. Ten marbles are randomly selected from Bowl 1 and transferred to Bowl 2. Then ten marbles are randomly selected from Bowl 2 and transferred to Bowl 1. Use algebra to show that there are as many white marbles in Bowl 2 as there are black marbles in Bowl 2.

A PERMUTATION TRICK

Overview of the Trick

The Trick

Ask a volunteer to secretly select two different single digits. Have them write each digit on its own index card and a decimal point on a third card, and then secretly construct all six decimals that can be formed by these three cards. Ask them to sum these six numbers.

Subsequently ask them to perform the following computations: multiply the sum by 100, divide the result by 11, and divide this result by 3. Finally, ask them to divide this last result by the sum of the two original digits selected. You immediately divine the result of their calculations is 37.

Example

Suppose 3 and 5 were chosen as the digits. Then the six possible decimal digits that are formed are:

53.   35.   5.3   3.5   .53   .35

The sum of these numbers is:

$53 + 35 + 5.3 + 3.5 + .53 + .35 = 97.68$

The subsequent calculations are:

$$97.68 \times 100 = 9768$$

$$9768 \div 11 = 888$$

$$888 \div 3 = 296$$

$$296 \div (3 + 5) = 37$$

Presto!

How it Works

This is an original trick. It relies on properties of permutations.

Main Concept

Polynomial factoring

Required Skills

Decimals
Decimal computations
Expanded notions for integers
Multiples
Distributive property
Use of variables
Use of algebraic expressions
Polynomials
Polynomial simplification
Polynomial addition
Factoring polynomials

Teaching and Performing the Trick

Suggested Modes of Instruction

Direct teaching, discovery learning, whole class, or cooperative groups

Materials Required

Index cards or other small scraps, paper and pencils are required. Calculators to check computations are optional.

Performing the Trick – Cooperative groups illustration

1. Before beginning the trick, secretly write the number 37 on several sheets of paper and seal each in an envelope.
2. Divide the class into several cooperative groups. Ask a volunteer from each group to secretly choose two different single-digit numbers. Have the volunteer secretly write each digit on an index card you supply and write a decimal point on a third card.
3. Give a volunteer from each group one of the envelopes to hold. Then have the volunteer show the other members of their group his or her choices.
4. Ask each group to construct all six decimal numbers that can be formed by these three cards. Then ask them to complete the following computations:

A PERMUTATION TRICK

a. Find the sum of the six decimals that were constructed.

b. Multiply the sum by 100.

c. Divide the product by 11.

d. Divide the quotient by 3.

e. Divide the new quotient by the sum of the original digits.

5. Feigning deep concentration, announce that you are predicting the answer to their computations and the correct answer will be sealed in the envelopes you gave them.
6. Distribute the *Student Activity Page* and work through it with the students.

Analysis – Using algebra to explain why it works

Represent the two chosen digits by a and b. The six possible arrangements as they would be written in decimal notation are:

ab. *ba.* *a.b* *b.a* *.ab* *.ba*

Summing these numbers and then writing the result in expanded notation gives:

$(a + b) \times 10 + (2a + 2b) \times 1 + (2a + 2b) \times {}^{1}/_{10} + (a + b) \times {}^{1}/_{100}$

After factoring a 2 from the expression $(2a + 2b)$, we have:

$(a + b) \times 10 + (a + b) \times 2 + (a + b) \times {}^{2}/_{10} + (a + b) \times {}^{1}/_{100}$

Note that $(a + b)$ is a common factor in the four terms of this sum. Factoring $(a + b)$ we have:

$(a + b)(10 + 2 + {}^{2}/_{10} + {}^{1}/_{100}) = (12.21)(a + b)$

Now, according to the directions in the problem, we compute as follows:

$12.21(a + b) \times 100 = 1221(a + b)$

$1221(a + b) \div 11 = 111(a + b)$

$111(a + b) \div 3 = 37(a + b)$

$37(a + b) \div (a + b) = 37$, as desired.

Further Investigations

Students investigate ways to generalize this trick.

A PERMUTATION TRICK

Name:

Steps in the Trick

1. **Choose two different single-digit numbers writing each digit on its own index card and writing a decimal point on a third index card.**
2. **Construct all six decimal numbers that can be formed by these three cards.**
3. **Complete the following computations:**
   a. **Find the sum of the six decimals that were constructed.**
   b. **Multiply the sum by 100.**
   c. **Divide the product by 11.**
   d. **Divide the quotient by 3.**
   e. **Divide the new quotient by the sum of the original digits.**
4. **You should be able to predict the result of these computations independent of the digits originally chosen.**

Unraveling the Trick

1. Choose two different variables to represent the two digits on the cards.
2. Write all the permutations of the three cards as decimals using the variables chosen.
3. Sum the permutations and express the sum in expanded notation.
4. Multiply the sum by **100** and express the result in factored form.
5. Explain the trick by showing that the algebraic expression from **4** is evenly divisible by **11**, **3**, and the sum of the two digits.

Extending Questions

1. Try the trick when the two selected digits are the same. Does it work?
2. Why was the sum multiplied by **100**? Could you predict the answer if this step was omitted?
3. Could one of the chosen digits be zero?

Further Investigations

1. Construct a variation of this trick using the following four cards: two cards with different single digits of one's choice, one card with a zero, and one card with a decimal point. How would you predict the answer to this trick?
2. Think of a new set of directions to be used (after the numbers have been summed) in the original trick. In this case how would you predict the answer?
3. Construct a trick using four cards: three with different single digits of one's choice and one with a decimal point. How would you predict the answer to this trick?
4. Analyze the following trick which also has **37** as the predicted answer.

   Choose any three-digit number with three identical digits. Divide the number by the sum of the digits. The result will always be **37**.

PERMUTATION PREDICTION

Overview of the Trick

The Trick

Ask a volunteer to secretly think of the three-digit number whose digits are in consecutive ascending order. Ask the volunteer to write down all possible arrangements of these three digits to form three-digit numbers–they should find six arrangements including the original number.

Ask the volunteer to add these numbers and then divide the result by 6. When the volunteer reveals their quotient you immediately discern their original number.

Example

If the number selected is 345, the permutations are: 345, 354, 435, 453, 534, and 543. The volunteer completes the computations below:

$$\begin{array}{r} 345 \\ 354 \\ 435 \\ 453 \\ 534 \\ +\underline{543} \\ 2664 \end{array}$$

$$\begin{array}{r} 444 \\ 6\overline{)2664} \end{array}$$

When the volunteer reveals the quotient, the repeated digit is the ten's digit of the original number, the hundreds and unit's digit one less ($4 - 1 = 3$) and one more ($4 + 1 = 5$), respectively. So you can announce that the original number was 345.

How it Works

This is a well-known trick which has appeared in mathematics texts and journals since the nineteenth century, all of the digits in the resulting quotient will be identical. Moreover, these digits are the ten's digit of the original number, making the hundred's and unit's digit of the original number one less and one more, respectively.

Main Concept

Factoring polynomials

Required Skills

Basic computation
Expanded notation for integers
Distributive property
Use of variables
Use of algebraic expressions
Polynomials
Polynomial simplification
Factoring polynomials

Teaching and Performing the Trick

Suggested Modes of Instruction

Direct teaching, whole class, or small groups

Materials Required

Paper and pencils are required. Calculators to check the results of these computations are optional.

Performing the Trick – Small group illustration

1. Divide the audience into small groups. Choose one volunteer from each group and briefly accompany this group of apprentice mathemagicians to a location away from the audience. Explain to them that they will be able to divine the three-digit number that members of their group choose following a sequence of calculations.

   Explain that the three-digit number must have digits in consecutive ascending order. The result of the calculations will be a three-digit number, all of whose digits are the same and are the middle digit of the number chosen by the group. They need only add 1 and subtract 1 to find the other digits. Make sure the apprentices understand these instructions. Return the apprentices to their groups.

PERMUTATION PREDICTION

2. Ask each group to secretly perform the following steps:
   a. Write down a three-digit number whose digits are in consecutive ascending order.
   b. Have the groups secretly write down all possible arrangements (permutations) of these three digits to form three-digit numbers. Each should find six arrangements including the original number.
   c. Add these six permutations of the original number and then divide the result by 6.
3. Ask the members of the group to reveal their quotient to the apprentice mathemagicians. They should be able to predict the original three-digit number.
4. If the students in different groups compare their results they will begin to see the pattern. Distribute the *Student Activity Page* and have it help them work cooperatively in groups to discover the algebraic mechanisms that underlie this trick.

Analysis – Using algebra to explain why it works

Before giving the algebraic explanation, we note that a proof by exhaustion is fairly simple to employ. For there are only 7 possible starting numbers: 123, 234, 345, 456, 567, 678, and 789. You can verify by computation that each of these 7 possible starting numbers will yield an answer with 3 repeated digits–always the ten's digit of the original number.

Since the three digits of the chosen number must be consecutive, represent them by a, $a + 1$, and $a + 2$. Note that $1 < a < 7$ (because 789 is the largest number you can use). In terms of a, the six permutations of the original number are listed in the table below.

	Hundred's place	Ten's place	Unit's place
The six permutations	a	$a + 1$	$a + 2$
	a	$a + 2$	$a + 1$
	$a + 1$	a	$a + 2$
	$a + 1$	$a + 2$	a
	$a + 2$	a	$a + 1$
	$a + 2$	$a + 1$	a
Column sums	$6a + 6$	$6a + 6$	$6a + 6$

The sum of the six permutations in expanded notation is:

$S = 100(6a + 6) + 10(6a + 6) + (6a + 6) =$
$600(a + 1) + 60(a + 1) + 6(a + 1)$

Now, dividing the sum, S, by 6 you get:

$S/_6 = 100(a + 1) + 10(a + 1) + (a + 1)$

This means that the answer will have $(a + 1)$ as a repeated digit. From this you can predict the three digits of the original number: a, $a + 1$, and $a + 2$.

Further Investigations

The penultimate division can be done in several different ways to reveal one of the original digits and hence the original number. One may investigate whether this trick will work with fewer or more than three digits in the original number.

PERMUTATION PREDICTION

Name:

Steps in the Trick

1. Choose any three-digit number whose digits are in consecutive ascending order.
2. Write down all possible arrangements (permutations) of these three digits to form three-digit numbers and list these six numbers.
3. Compute the sum of these numbers and then divide the sum by 6.
4. From the quotient, the original number can be predicted.

Unraveling the Trick

1. Using a single variable, write an algebraic expression for each digit in the original number starting with the hundred's digit.
2. Write the six possible arrangements for each of the three expressions you just found. Line up the expressions for the six permutations in three columns.
3. Find the sum of each column and divide each sum by **6**.
4. Use the results of the last step to explain the trick.

Extending Questions

1. What are the possible values for your chosen variable?
2. Would the trick work if you divide the sum by **3** and then followed that by a division of **2**?
3. What is the smallest and largest number that we can use for the last digit in the chosen number?
4. What is the smallest and largest number that we can use for the first digit in the chosen number?
5. When Unraveling the Trick, try using $a - 1$, a, and $a + 1$ to represent the digits in the number? Can you predict the original number? Could you use $a + 2$, $a + 3$, and $a + 4$?

Further Investigations

1. What would happen if you choose three consecutive even (or odd) digits? Why does or doesn't this work?
2. Here is a variation for the ending of the trick. Consider adding the following two computations after completing the division by **6**. Divide the quotient by **3** and then divide the new quotient by **37**. Experiment to see the effect of these computations on the trick. How does each step affect the result? Can you predict the original number from the final answer? Why or why not?
3. Here is another possible way to end the trick. After the original two computations, ask the subject to divide the quotient by the sum of the three original digits. Is this result predictable? Why or why not?
4. Can you adapt this trick to work for two-digit numbers? Is it much of a trick then?
5. Examine what happens when a three-digit number is replaced by a four-digit number. In this case, can you find a pattern in the answers which allows you to predict the original number?

DIVINING TWO NUMBERS

Overview of the Trick

The Trick

A volunteer is asked to choose two single-digit numbers. Ask them to mentally perform a sequence of arithmetical calculations based on their numbers: multiply the first number by 2; add 3 ; multiply by 5; add the second number; multiply by 10. Asking the volunteer to reveal the result of the computation, you immediately reveal the two numbers chosen by the volunteer.

Example

Suppose a 5 and a 3 were chosen, the computations are as follow:

$$2 \times 5 = 10$$
$$10 + 3 = 13$$
$$5 \times 13 = 65$$
$$65 + 3 = 68$$
$$68 \times 10 = 680$$

When 150 is subtracted from 680, the result (530) reveals that the subject chose 5 as their first number and 3 as their second number.

How it Works

Every magician knows tricks to predict (or divine) more than one number thought of by a volunteer. As presented here, this trick works with single-digit numbers which can be generated by dice rolls, domino draws, or card cuts. The trick works by secretly encoding the first number in the hundred's digit and the second number in the ten's digit of the computation the volunteer performs.

Main Concept

Factoring polynomials

Required Skills

Basic computation
Expanded notation of integers
Distributive property
Use of variables
Use of algebraic expressions
Polynomials
Polynomial addition
Factoring polynomials

Teaching and Performing the Trick

Suggested Modes of Instruction

Direct teaching, whole class, small groups, or discovery learning

Materials Required

Regular six-sided dice, standard dominoes, or a deck of cards with 10s and face cards removed can all be used to generate the two mystery numbers are optional.

Performing the Trick – Discovery learning illustration

1. Using dice, dominoes, playing cards, or some other manipulative, choose two single-digit numbers out of sight of the audience.
2. Tell the audience you will mentally perform a sequence of arithmetical computations on these two numbers. If they follow along algebraically, calling the two numbers you have chosen x and y, they might learn the magic secret of divining two numbers.
3. Mentally perform the following computations, keeping the results of the computations secret as you proceed, and allowing the audience enough time to perform the corresponding algebraic computations:
   a. Multiply the first number by 2.
   b. Add 3 to this number.
   c. Multiply the result by 5.
   d. Add the second number to this product.
   e. Multiply the sum by 10.

DIVINING TWO NUMBERS

4. Announce the final result of your computation.

5. Wait. If members of the audience have completed their algebraic computations correctly, they will be able to solve a simple algebraic equation to divine your numbers. (The result of your computation will be equal to $100x + 10y + 150$ which can be readily solved.)

Analysis – Using algebra to explain why it works

Let x and y be the first and second numbers chosen, respectively. Then the steps in the computation from the second stage of the trick are as follow:

Multiply the first number by 2.	$2x$
Add 3 to the result.	$2x + 3$
Multiply the result by 5.	$5(2x + 3) = 10x + 15$
Add the second number to the result.	$10x + 15 + y$
Multiply the resulting sum by 10.	$10(10x + 15 + y) = 100x + 10y + 150$

When the last number is reported, subtract 150, and the result is always $100x + 10y + 0$. Thus, x and y are the digits in the hundred's and ten's places, allowing you to announce them to your subject. A zero will always be in the unit's place.

Further Investigations

Students investigate ways to generalize this trick: divining multi-digit numbers and/or more than single-digit numbers.

DIVINING TWO NUMBERS

Name:

Steps in the Trick

1. Have a volunteer choose two single-digit numbers.
2. Ask the volunteer to mentally perform the following computations, keeping their computations secret as they proceed:
   a. Multiply the first number by 2.
   b. Add 3 to this number.
   c. Multiply the result by 5.
   d. Add the second number to this product.
   e. Multiply the sum by 10.
3. Have the volunteer announce the final result of the computation.
4. From this result, you magically divine the two chosen numbers.

Unraveling the Trick

1. Use two different variables to represent the two chosen mystery numbers.
2. Write algebraic expressions to represent the resulting numbers after each computational step in the trick.
3. Subtract **150** from the last algebraic expression in **2** and use this result to explain the trick.

Extending Questions

1. If the two single-digit numbers were identical, could you still predict the numbers?
2. If the subject chooses two numbers one of which is zero, will the trick still work?
3. If the first number chosen was **10**, **11**, or **12** and the second a single-digit number, can you still make a prediction?
4. Could you use dodecahedra (12-sided) or icosahedra (20-sided) dice for the trick?

Further Investigations

1. Use algebra to determine how to complete the steps of the following trick to divine three single-digit numbers:
   a. Double the first number.
   b. Add **1** to this product.
   c. Multiply this sum by **5**.
   d. Add the second number to this product.
   e. Double this sum.
   f. Add **1** to this product.
   g. Multiply this sum by **5**.
2. Develop a trick which divines four or more single-digit numbers.
3. Choose two two-digit numbers. Starting with one of these numbers, perform **Steps a**, **b**, **c**, **e**, **f**, and then **g** of the trick in the first investigation in this section. Explain how you can complete these steps to divine the two original numbers.
4. Algebraically analyze your trick from the previous investigation.

THE MISSING DIGIT

Overview of the Trick

The Trick

Ask a volunteer to secretly select a four-digit number. He/she is asked to compute the sum of the digits of this number. He/she then crosses out one of the non-zero digits in the original number to form a three-digit number. From this three-digit number they subtract the sum of the original digits and then compute the sum of the digits in the resulting difference. When they announce this difference, you immediately divine the digit that vanished.

Example

Suppose the number chosen is 3672. The sum of the digits is $3 + 6 + 7 + 2 = 18$. If the volunteer deletes the 7 then they form the three-digit number 362 and compute $362 - 18 = 344$. The sum of these digits is $3 + 4 + 4 = 11$. When this sum is revealed you mentally determine that it is 7 less than the next smallest multiple of 9 which is bigger than the announced sum, so 7 must be the deleted digit.

How it Works

Like *Trick 18–Magical Number Nine*, this trick uses special properties of the number 9. Here it uses the fact that the difference of any number and the sum of the digits of that number is always a multiple of 9.

Main Concept

Factoring polynomials

Required Skills

Basic computation
Expanded notation for integers
Distributive property
Use of variables
Use of algebraic expressions
Polynomials
Simplifying polynomials

Teaching and Performing the Trick

Suggested Modes of Instruction

Direct teaching, small groups, or whole class

Materials Required

Paper and pencils for computations are required. Calculators to check the results of computations are optional.

Performing the Trick – Whole class illustration

1. With the audience together, ask each member to secretly select a four-digit number.
2. Have each member determine the sum of the digits of their number.
3. Delete any one of the non-zero digits in their original number and write the remaining three digits as a three-digit number.
4. From the three-digit number, subtract the sum of the digits of the original number.
5. Sum the digits in this difference.
6. Randomly ask volunteers to reveal their resulting sum. Subtract their number from the smallest multiple of nine bigger than this number. The result will be the number they deleted as long as it was not a 9. If their final number is already a multiple of 9, the deleted digit is a 9.
7. Encourage students to try to find a pattern between their announced numbers and the missing digit. Distribute the *Student Activity Page* and have them work cooperatively through them in groups.

Analysis – Using algebra to explain why it works

The key to this trick is that the difference of any number and the sum of its digits is divisible by 9. To see this is true for a four-digit number, suppose the four-digit number is written $N = abcd$.

THE MISSING DIGIT

Written in expanded notation, we have:

$N = 1000a + 100b + 10c + d$

The sum of the digits of N is $S = (a + b + c + d)$. Hence, the difference between the number and the sum is its digits is:

$$N - S = (1000a + 100b + 10c + d) - (a + b + c + d)$$
$$= 999a + 99b + 9c$$
$$= 9(111a + 11b + c)$$

The result is a multiple of 9.

For this specific trick, let the chosen number be written as $N = abcd$ as above. If we delete the digit c in N, then the resulting three-digit number without the digit c is given by $100a + 10b + d$ in expanded notation. In the trick the volunteer subtracts the sum of the original digits from this three-digit number, giving:

$$100a + 10b + d - (a + b + c + d) = 99a + 9b - c$$
$$= 9(11a + b) - c$$

The result is a multiple of 9 minus c. When this number is reported, you need only determine what one-digit number must be added to the result to produce a multiple of 9. The number you add will be the same as the missing digit, c.

The argument will be similar no matter which of the digits (a, b, c, or d) is crossed out.

Further Investigations

Students investigate both generalizations and limitations of this trick.

THE MISSING DIGIT

Name:

Steps in the Trick

1. Choose a four-digit number and compute the sum of its digits.
2. Delete any one of the non-zero digits in the original number and write the remaining three digits as a three-digit number.
3. From the three-digit number, subtract the sum of the digits of the original number.
4. Sum the digits in this difference.
5. From the resulting sum you can predict the deleted digit.

Unraveling the Trick

1. Choose four different variables to represent the thousand's, hundred's, ten's and unit's digits of the four digit number, N. Express N in expanded notation.
2. Represent the sum of the four digits of N by S.
3. Use algebra to show that $(N - S)$ is a multiple of **9**.
4. Delete one digit from the four-digit number N and express the remaining three digits as a three-digit number in expanded notation. Represent this number by M.
5. Find $(M - S)$ and simplify the result.
6. Explain the trick by using the observation from **3** and considering what number must be added to the expression in **5** to produce a multiple of **9**.

Extending Questions

1. Try the trick if all four digits are identical. Can you still predict the deleted digit?
2. Would the trick work if a **0** were deleted from the original number? Try it.
3. Try the trick by deleting a **9** from the chosen number. Can you predict the deleted digit?
4. Based on Questions 2 and 3 above, how could you alter the directions so the trick will always work?

Further Investigations

1. Verify the argument in ***Unraveling the Trick*** if a different digit from the one you chose previously is deleted from the number, N.
2. Why must the deleted digit be non-zero?
3. Provide numerical examples to show that this trick works equally well with three-digit numbers and six-digit numbers.
4. Use algebra to verify that the following variation of the trick always works.

   Ask a volunteer to write down a three-digit number and then reverse the digits to form another three-digit number. Subtract these two numbers, the smaller from the larger. If the volunteer tells you the number of digits in the remainder as well as the left-hand digit in the remainder, you can tell the whole remainder. If there are three digits, the middle digit is always a ***9****. In either case the left-hand and right-hand digits sum to* ***9****. Hence, having the left-hand digit (say* ***2****), the whole remainder would be known (****297*** *in this case).*
5. Give numerical illustrations of the following trick and determine what happens if **0** or **9** is deleted from the chosen number.

   Ask a volunteer to think of a number of three or more digits. Have them divide this number by ***9*** *and reveal the remainder. Delete one digit from the original number, divide this new number by* ***9****, and reveal this remainder as well. You can tell what digit was deleted. The trick is this: If the second remainder is less than the first, the figure erased is the difference between the remainders. If the second remainder is greater than the first, the digit is* ***9*** *minus the difference of the remainders.*
6. Use algebra to verify the following trick. Write down any number. Find the sum of its digits, multiply by **8**, and add the result to the original number. The result is a multiple of **9**, that is, the remainder is **0** when divided by nine.

THINK OF A NUMBER

Overview of the Trick

The Trick

Ask a volunteer to secretly select a number. Then ask the volunteer to perform a sequence of calculations: Add 1 to the number chosen. Multiply the sum by 3. Add the square of the original number to this product. Multiply this sum by 4. Lastly, subtract 3 from the last product and take the square root of the difference. When the result of these computations is announced, you correctly divine the identity of their original number.

Example

Suppose the original number chosen was 7. Then the computations are as follows:

$$7 + 1 = 8$$
$$8 \times 3 = 24$$
$$24 + 7^2 = 73$$
$$73 \times 4 = 292$$
$$292 - 3 = 289$$
$$\sqrt{289} = 17$$

When the volunteer announces the result of the computation, you mentally subtract 3 and divide by 2 to divine their original number.

How it Works

Tricks like this are one of the oldest and most used number tricks. Tricks like this appeared in early texts on number games and the word "divining" was commonly used. Such tricks have now found their way to the Internet where people often circulate "amazing" emails that predict anything from the year you were born to the number of times a day you would like to eat chocolate.

All of these tricks rely on algebraic identities which simply camouflage the original number.

Main Concepts

Multiplying polynomials, factoring polynomials, and solving equations

Required Skills

Basic computation
Distributive property
Use of variables
Use of algebraic expressions
Polynomials
Simplifying polynomials
Adding polynomials
Multiplying polynomials
Solving equation

Teaching and Performing the Trick

Suggested Modes of Instruction

Direct teaching, whole class, discovery learning, or small groups

Materials Required

Paper and pencils. Calculators to check computations are optional.

Performing the Trick – Small groups illustration

1. Divide the audience into small groups. Chose one volunteer from each group and briefly accompany this group of apprentice mathemagicians to a location away from the audience. Explain to them that they will be able to divine the numbers originally chosen by the members of their group by subtracting 3 from the result of each person's computation and then dividing by 2. Be sure they are comfortable doing this mentally or that they are equipped with a calculator. Return the apprentices to their groups.
2. Ask each member of the audience, save the apprentice mathemagicians, to secretly select a number.
3. Ask them to complete the following computations, keeping these computations secret from the apprentice mathemagicians at their table:

THINK OF A NUMBER

a. Add 1 to the original number.

b. Multiply this sum by 3.

c. Add the square of the original number to this product.

d. Multiply this sum by 4.

e. Subtract 3 from this product.

f. Take the square root of the difference.

4. Ask each the members of the audience to reveal the results of their computations to the apprentice mathemagicians, one at a time, so the apprentices can divine the numbers they thought of.

5. Circulate among the groups to check on the progress of your apprentices and deal with any miscalculations that might have caused difficulties. Allow the apprentices to reveal their "trick" after all of the numbers have been divined.

6. Distribute *Student Activity Page* and have the groups work through them cooperatively, electing other members of their group to serve as apprentices for the alternative tricks that are part of these pages.

Analysis – Using algebra to explain why it works

Denote the number selected by a volunteer by x. The algebraic computations that correspond to the arithmetical computations performed by the volunteer are as follows:

Add 1 to the original number.	$x + 1$
Multiply the sum by 3.	$3(x + 1) = 3x + 3$
Add the square of the original number to the product.	$x^2 + (3x + 3) = x^2 + 3x + 3$
Multiply the sum by 4.	$4(x^2 + 3x + 3) = 4x^2 + 12x + 12$
Subtract 3 from the product.	$4x^2 + 12x + 12 - 3 = 4x^2 + 12x + 9 = (2x + 3)^2$
Take the square root of the difference.	$\sqrt{(2x + 3)^2} = 2x + 3$

Thus, when 3 is subtracted from the answer reported by the volunteer the result is $2x$ and hence, division by 2 correctly divines the original number x.

Such number tricks can be constructed from any algebraic identity. In practice, it is a good idea to vary the trick unless your goal is for the audience to figure out the pattern. In general it is best to tailor the trick to suit the level of arithmetical/algebraic facility of the audience.

When trying to impress upon an audience that such tricks are no more than algebraic identities at work, the following "trick" based on the identity $(x + 3) - 3 = x$ is often helpful:

1. Choose a number.
2. Add 3.
3. Tell me the result of your calculation.
4. I bet your number was _____. Isn't that amazing!

Further Investigations

Students are asked to investigate similar computational tricks and then use algebraic identities to create their own computational tricks.

THINK OF A NUMBER

Name:

Steps in the Trick

1. Choose a number.
2. Complete the following computations:
   a. Add 1 to the original number.
   b. Multiply this sum by 3.
   c. Add the square of the original number to this product.
   d. Multiply this sum by 4.
   e. Subtract 3 from this product.
   f. Take the square root of the difference.
3. From the result of these computations you can predict ("divine") the identity of the original number.

Unraveling the Trick

1. Choose a variable to represent the selected number.
2. Write algebraic expressions for the number after each computational step in **Step 2** of the trick.
3. Simplify the last algebraic expression and subtract **3** from it and divide the difference by **2**.
4. Explain the trick by comparing the result of **3** with the original number chosen.

Extending Questions

1. Try starting the trick with **–5**. Try starting with **0**. Does the number chosen have to be positive for the trick to work? Explain.
2. Try the trick with several fractional values for the number chosen. Will this trick work for any positive rational number? Explain.
3. Is $\sqrt{(2x + 3)^2} = 2x + 3$ always true? [Hint: try $x = -3$.]
4. Alter the directions for the trick to take the observations in Questions 1–3 above into account.

Further Investigations

1. Three tricks are show below. Show why these tricks always work.

Trick 1

- Think of a number, say **6**.
- Multiply this number by **3**. $6 \times 3 = 18$
- Add **1** to the product. $18 + 1 = 19$
- Multiply the sum by **3**. $19 \times 3 = 57$
- Add the product to the original number. $57 + 6 = 63$
- Report the result.

The reported number will always end with **3**. Delete the **3**. What remains is the original number (**6** in this example).

Trick 2

- Think of a number, say **6**.
- Double this number. $6 \times 2 = 12$
- Add **4** to the product. $12 + 4 = 16$
- Multiply the sum by **5**. $16 \times 5 = 80$
- Add **12** to the product. $80 + 12 = 92$
- Multiply the sum by **10**. $92 \times 10 = 920$
- Report the result.

Subtract **320** from the reported number (**920 – 320 = 600** in this case). Delete the zeroes in the ten's and unit's digits. The resulting number is the original number (**6**).

THINK OF A NUMBER

Name:

Trick 3

- Think of a number, say **6**.
- Multiply this number by itself. **6 × 6 = 36**
- Subtract **1** from the original number. **6 – 1 = 5**
- Multiply the difference by itself. **5 × 5 = 25**
- Find the difference between the earlier product and this last product. **36 – 25 = 11**
- Report the result.

Add **1** to the reported number (**11 + 1 = 12** in this case) and divide this number by **2** (**12 ÷ 2 = 6**). This gives the original number (**6**).

2. Any algebraic identity may be used as the basis of a number trick of number game. For example, the equation $\sqrt{\frac{(5x)^2 + (12x)^2}{x^2}} = 13$

   leads to the following trick:

   Think of a number. First multiply your number by ***5*** *and then square the result. After that, multiply your original number by* ***12*** *and square the result. Add these two squares together and divide their sum by the square of the original number. Compute the square root of this difference (using a calculator if necessary). Don't tell me the answer, for I already know what it is:* ***13***.

   Use this example to devise a trick for the equation $x^2 + 2x + 1 = (x + 1)^2$.

3. Devise a trick for the equation

$$\frac{\sqrt{(3x)^2 + (4x)^2} + 15}{5} = x + 3$$

4. Find other algebraic identities and write divining tricks based on them.

MAGICAL NUMBER NINE

Overview of the Trick

The Trick

Ask a volunteer to secretly select a three-digit number with no repeating digits. Then ask them to form another three-digit number by rearranging the original digits. Ask them to take the difference of these two numbers and add the digits of this difference together. When the volunteer reveals whether the sum of the digits contains one or two digits, you immediately predict what this sum is.

Example

Suppose the number chosen is 357 and 735 is chosen as the permutation of the digits. The volunteer then computes the difference: $735 - 357 = 378$ and then the sum of the digits: $3 + 7 + 8 = 18$. When the volunteer reveals that the sum of the digits is a two digit number you reveal that this sum must be 18.

Alternatively, suppose the number chosen is 394 and the permutation is 349. Then the difference is $394 - 349 = 45$ and the sum of the digits is simply 9. When the volunteer reveals that the sum of the digits is a two-digit number you reveal that the sum must be 9.

How it Works

"Casting out nines" uses familiar properties of the number nine to check arithmetical computations. These same properties are the basis for numerous number tricks. In this trick the sum of the digits must always be either 9 or 18. Once you know whether the sum has one- or two-digits, you know what this sum must be.

Main Concept

Polynomial addition

Required Skills

Basic computation
Divisibility properties
Permutations
Expanded notation
Multiples
Distributive property
Use of variables
Use of algebraic expressions
Polynomials
Simplifying polynomials
Polynomial addition

Teaching and Performing the Trick

Suggested Modes of Instruction

Direct teaching, small groups, or whole class

Materials Required

Paper and pencils are required. Calculators to check computations are optional.

Performing the Trick – Whole class illustration

1. With the audience together, ask each member to secretly select a three-digit number with no repeating digits.
2. Have each member form another three-digit number by rearranging (permuting) the digits in their original number.
3. Find the difference between these two numbers, subtracting the smaller from the larger.
4. Add the digits of this difference together.
5. Randomly ask volunteers to reveal the number of digits in their resulting sum. There will either be one or two digits. If the number of digits is one, their sum was 9. If the number of digits was two, their sum was 18.
6. Students will begin to see the pattern. Distribute the *Student Activity Page* and work through it with the students.

MAGICAL NUMBER NINE

Analysis – Using algebra to explain why it works

Because the trick allows the subject to choose any permutation of the original three digits, the algebraic verification requires examining several possible cases. One case is illustrated below, you may wish to try others.

Suppose the three-digit number chosen is written *abc*, where none of a, b, or c are equal. Let us assume that $a > c$. Also, let us assume that the permutation of the digits considered is *cba*. To subtract *cba* from *abc* requires borrowing from the hundred's column and then from the ten's column as shown below.

100's	10's	unit's	100's	10's	unit's	100's	10's	unit's
a	b	c	$a-1$	$b+10$	c	$a-1$	$b+9$	$c+10$
$-c$	b	a	$-c$	b	a	$-c$	b	a
						$(a-1-c)$	$(b+9-b)$	$(c+10-a)$

Writing this difference in expanded notation we get:

$100(a-1-c) + 10(b+9-b) + (c+10-a) = 99a - 99c$
$= 9(11a - 11c)$

which is a multiple of 9. The sum of these digits, $(a-1-c) + (b+9-b) + (c+10-a) = 18$

You may wish to verify other cases for this trick. In each the sum of the digits is either 9 or 18.

More generally, when we use a different number of digits than three, what we can say about the sum of the digits of the difference is that it is always a multiple of nine. Hence, knowing some information about the sum of the digits allows you to divine other information. For example, if you know the sum of the digits is a two-digit number with unit's digit 4 the sum must be 54 as this is the only two-digit multiple of 9 ending in 4.

In using three-digit numbers the difference must be three-digits or fewer so the sum of the digits is strictly less than 27, leaving 18, 9, and 0 as the only possible sums for the digits of the difference. 0 may occur if there are repeated digits.

Further Investigations

This trick can be adapted in many ways depending on the number of digits in the original numbers. Students are asked to investigate several different variations.

MAGICAL NUMBER NINE

Name:

Steps in the Trick

1. **Choose a three-digit number with no digits repeating.**
2. **Write down another three-digit number by rearranging the digits in their original number.**
3. **Find the difference between these two numbers, subtracting the smaller from the larger.**
4. **Add the digits of this difference together.**
5. **From the number of digits in sum you can predict the exact value of this sum.**

Unraveling the Trick

1. Choose three different variables for the hundred's, ten's, and unit's digit of the number, and write them in appropriately labeled columns.
2. Borrow one from the hundred's column, and rewrite the hundred's and ten's digits appropriately.
3. Borrow one from the ten's column, and rewrite the ten's and unit's digits appropriately.
4. Suppose the permutation selected is the original digits in reverse order. Subtract these digits from those appearing in the table.
5. Sum the digits in the difference above, and use this result to explain the trick.

Extending Questions

1. Try the trick when all of the digits in the chosen number are the same. What happens?
2. Try the trick when two of the digits are the same. Does the trick work?
3. Try the trick using a permutation other than the original digits reversed. Will the sum of the digits in the difference always be a multiple of **9**?

Further Investigations

1. Using algebra, verify the following variation of this trick.

   Think of a two-digit number with unequal digits. Reverse the digits and find the difference of the two numbers. Tell me one of the digits in the difference and I will tell you the other.

2. Show with numerical illustrations that the following variation of the trick works.

   *Take a two-digit number with unequal digits and transpose the digits. Square each of these numbers and compute the difference. Now, add the digits in the difference. The difference will be **18**.*

3. Using what you learned about the sum of the digits of the difference being a multiple of **9**, provide numerical illustrations for the following variation of this trick:

   Think of any number with unequal digits. Rearrange the digits to form a new number. Find the difference between these two numbers and the sum of the digits of the difference. Tell me all but one of the digits in the sum and I will tell you the missing digit.

DIVINING FROM A TABLE

Overview of the Trick

The Trick

Ask a volunteer to secretly select a number between 1 and 63. Give the volunteer six cards containing arrays of numbers. Ask the volunteer to give back to you each of the cards on which his/her number appears. You immediately divine the number they chose.

Example

Suppose the volunteer selects the number 50. Three cards contain the number 50. As they are returned to you, you simply add the upper left hand entries on each of these cards: 2 + 16 + 32 = 50. You will always arrive at the desired result.

How it Works

This trick is based on the binary representation of the number chosen. In fact, by giving you the cards that contain the original number, the volunteer has given you the binary representation. You simply convert this number back into its base ten representation by adding the binary terms.

Although we may associate binary arithmetic with the computer age, number systems in other bases have long been important. This trick was published as early as 1857 in *The magician's own book*, New York: Dick and Fitzgerald Publishers, p. 241.

Main Concept

Translation from base-two representations to base-ten representations

Required Skills

Basic computation
Multiples
Expanded notation

Teaching and Performing the Trick

Suggested Modes of Instruction

Direct teaching, whole class, or small groups

Materials

Copies of the binary cards that appear below are required. For repeated use they should be copied onto cardstock, laminated, or made as transparencies.

Performing the Trick – Small groups illustration

1. Divide the audience into small groups. Choose one volunteer from each group and briefly accompany this group of apprentice mathemagicians to an isolated location away from the audience. Explain to them that they will be able to divine the number chosen by the volunteer from their group by adding the upper right-hand entries on the cards that are returned to them by the volunteer. Be sure they are comfortable doing this mentally or that they are equipped with a calculator. Return the apprentices to their groups.
2. Ask each member of each group, save the apprentice mathemagicians, to secretly think of a number between 1 and 63. Have them share the identity of their number with the other members of the group save the apprentice.
3. Ask the volunteers to return to the apprentice each of the cards that contains their number.
4. The apprentices will divine the volunteer's chosen number by adding the upper left-hand digits on each of the cards that has been returned to them.
5. Circulate among the groups to check on the progress of your apprentices and deal with any difficulties. Have the groups choose another volunteer and repeat the trick. After a few examples, have the apprentices teach a new apprentice from the group how to perform the trick and give them an opportunity to divine several numbers.
6. Distribute *Student Activity Pages* and have the groups work through them cooperatively.

Analysis – Using algebra to explain why it works

Distribution of the numbers 1 to 63 on the six cards is based upon the binary coding of each number. All numbers that have a 1 in the unit's place of the binary code appear on the "1" card. All numbers that have a 1 in the two's place (second place from the right) of the binary code appear on the "2" card. All numbers that have a 1 in the four's place (third from

DIVINING FROM A TABLE

the right) in the binary code appear on the "4" card, etc.

For example, the binary code for 50 is 110 010, which in expanded notation is:

$1 \times 2^5 + 1 \times 2^4 + 0 \times 2^3 + 0 \times 2^2 + 1 \times 2^1 + 0 \times 2^0 =$
$1 \times 32 + 1 \times 16 + 0 \times 8 + 0 \times 4 + 1 \times 2 + 0 \times 1 = 50$

Thus, 50 appears on the "2" card, the "16" card, and on the "32" card. When these cards are returned to you, you need only add the values in the upper right corner, 2 + 16 + 32 = 50, to determine the chosen number.

In general, any number x from 1 to 63 has a unique binary code consisting of a string of 0's and 1's no longer than six digits. When this binary code is written in expanded notation, x will be the sum of a unique combination of the following set of values: 1, 2, 4, 8, 16, 32. Another way of saying this is that x will appear on a unique combination of the six binary cards shown below and to the right. Thus, summing the values in the upper left-hand corners (the powers of two) will produce the value x.

1	3	5	7
9	11	13	15
17	19	21	23
25	27	29	31
33	35	37	39
41	43	45	47
49	51	53	55
57	59	61	63

2	3	6	7
10	11	14	15
18	19	22	23
26	27	30	31
34	35	38	39
42	43	46	47
50	51	54	55
58	59	62	63

4	5	6	7
12	13	14	15
20	21	22	23
28	29	30	31
36	37	38	39
44	45	46	47
52	53	54	55
60	61	62	63

8	9	10	11
12	13	14	15
24	25	26	27
28	29	30	31
40	41	42	43
44	45	46	47
56	57	58	59
60	61	62	63

16	17	18	19
20	21	22	23
24	25	26	27
28	29	30	31
48	49	50	51
52	53	54	55
56	57	58	59
60	61	62	63

32	33	34	35
36	37	38	39
40	41	42	43
44	45	46	47
48	49	50	51
52	53	54	55
56	57	58	59
60	61	62	63

Further Investigations

Students investigate generalizations of this trick to include larger numbers and adapting this trick to other bases.

DIVINING FROM A TABLE

Name:

Steps in the Trick

1. Ask a volunteer to secretly think of a number between 1 and 64.
2. Give the volunteer the six cards.
3. Ask the volunteer to return to you each of the cards that contains his/her number.
4. You will be able to immediately divine their secret number.

Unraveling the Trick

1. Verify that all the numbers on the **1 card** (the card with a **1** in the upper left-hand corner) have a **1** in the unit's place of their binary codes.
2. Verify that all numbers on the **2 card** have a **1** in the two's place of their binary codes.
3. Verify that all the numbers on the **4**, **8**, **16**, and **32 cards** have **1**s in the four's, eight's, sixteen's, and thirty-two's place (respectively) of their binary codes.
4. Use the observations above to explain why any number from **1** to **63** written in base ten can be predicted by adding the numbers in the upper right corner of each card on which that given number appears.

Extending Questions

1. With reference to the binary system, what is the significance of the set of numbers found in the upper left-hand corners of the six cards?
2. Why is the binary system especially well-suited for this trick?
3. What is the sum of the six numbers found in the upper left-hand corners of the cards?
4. If a seventh card was constructed, what number would you place in the upper right corner? What is the largest number that could be predicted using a set of seven binary cards?

Further Investigations

1. How many binary cards would be needed to allow a subject to think of a number up to **500**?
2. If you constructed **8** binary cards, how large a number could you predict? Suppose you had ***n*** binary cards?
3. The following computations illustrate an algorithm for changing the decimal number, **50**, to its binary code, **110 010**.

Computation	Quotient	Remainder	Binary coding
50 ÷ 2 =	25	0	unit's place
25 ÷ 2 =	12	1	two's place
12 ÷ 2 =	6	0	four's place
6 ÷ 2 =	3	0	eight's place
3 ÷ 2 =	1	1	sixteen's place
1 ÷ 2 =	0	1	thirty-two's place

Thus, **50** in base ten is **110 010** in base two.

Use this algorithm to find the binary code for other numbers between **1** and **64**. Verify your coding by checking the six binary cards. If you have a scientific calculator which converts base ten to base two, use this to verify your codes.

4. Explain why the algorithm above works.
5. Devise a set of cards based on ternary coding (base three notation) that could be used to predict a number.

THE DOMINO CHAIN

Overview of the Trick

The Trick

Ask a volunteer to make a single chain using all of the dominoes in a set of double-six dominoes. When the volunteer finishes the chain, you have them open a sealed envelope you gave them at the outset of the trick which correctly predicts the beginning and ending numbers in their chain of dominoes.

Example

Before the trick you secretly remove one domino from the set; suppose you removed the 3–5 domino. In an envelope, you seal your prediction *3 and 5*. No matter how your subject begins the chain, to use all the pieces on the table, the completed chain will begin and end with the numbers 3 and 5.

How it Works

This trick uses sleight of hand. Before you mix and place the dominoes face up on the table, you must remove and conceal one piece. Make sure the domino you conceal is not a double. Your prediction will consist of the two numbers on this hidden domino. When your subject has completed the chain using all of the dominoes on the table, the first and last numbers in the chain will match your prediction.

If the trick is to be repeated, secretly return the piece to the set and remove a different piece. In this way, your prediction will vary each time you perform the trick.

Main Concept

Counting arguments

Required Skills

Basic computation
Use of variables
Algebraic expressions

Teaching and Performing the Trick

Suggested Modes of Instruction

Direct teaching, small groups, or cooperative groups

Materials Required

One or more complete sets of double-six, 28-piece domino sets are required.

Performing the Trick – Cooperative groups illustration

1. Prior to performing the trick, secretly remove one non-double domino from each of the complete sets. In sealed envelopes, write your prediction, the numbers on the removed domino, on a sheet of paper.
2. Divide the class into several groups. Give each group a set of dominoes and a sealed envelope. Make sure that the envelopes correspond to the single domino that was secretly removed from the group's set.
3. Ask each group to form a single chain using all of the dominoes in the set, matching the dominoes end to end as in normal play.
4. After the chain is completed, have the groups open the envelopes and check to see if your prediction was correct.
5. Have each group return the prediction to a new sealed envelope, mix up their domino chain, and then switch materials with another group. Repeat the trick.
6. As the students begin to realize there is something special about the various sets of dominoes, hand out the *Student Activity Page* and have them work through it in groups.

Analysis – Using algebra to explain why it works

In a standard double-six domino set there are 28 dominoes and each of the numbers 0–6 appears a total of eight times on a total of seven different tiles. Because each number appears an even number of times, in any complete chain using the whole set, the two end numbers on the chain have to match.

By secretly removing a domino that is not a double, you change the parity. Suppose you have removed the domino with the numbers x and y where $x \neq y$. The numbers x and y will only appear seven times on the remaining dominoes which the volunteer uses to make the chain. Because of this, every complete chain must begin with either x or y and end with the other.

Further Investigations

Students investigate whether and how this trick can be adapted for alternative sets of dominoes.

THE DOMINO CHAIN

Name:

Steps in the Trick

1. **Mix all but one of a 28-piece set of double-six dominoes and place them face up on a table.**
2. **Ask a volunteer to form a single chain using all of the dominoes, matching them end to end as in normal play.**
3. **You can predict the beginning and ending numbers in the domino chain.**

Unraveling the Trick

1. Using a double-six set of dominoes for a single chain that begins with the **4–5** domino. What is the ending number in the chain?
2. Now remove the **4–5** domino and make a single chain with the **27** remaining dominoes. What are the beginning and ending numbers in your completed chain?
3. Why does the removal of the **4–5** domino guarantee the result you found in the preceding question?
4. Remove the **5–5** domino from the set and form a chain with the remaining **27** dominoes. Does your chain begin and end with **5**? If it does, can you change it so it does not?

Further Investigations

1. Use a 28-piece set of dominoes to experiment with chains. Start a chain with a **3–3** domino and see if you can end with a number other than **3**. Start with a **3–*x*** domino (where ***x*** is not **3**) and see if you can end with a number other than **3**. Write an explanation of why any chain that begins with ***x*** must end with ***x***.
2. Explain why the trick does not work if you conceal a double domino.
3. A double-five set of dominoes (from **0–0** to **5–5**) has 21 pieces. Experiment with such a set to show that a complete chain of the 21 pieces does not have to begin and end with the same number. Can you form a complete chain using all 21 pieces?
4. Show that the number of dominoes in a set going from a **0–0** domino to an ***n–n*** domino is

$$\frac{[(n + 1)(n + 2)]}{2}$$

5. There are millions of ways that a 28-piece set of dominoes can be arranged in a chain using the rules of the game. Yet it is impossible to put a set double-five set of dominoes (21 pieces) into a continuous chain using all 21 pieces (see Investigation 3 above). If you removed the **0–2** and the **1–3** dominoes from a double-five set, however, you can form a complete chain with the remaining 19 pieces. Try it.

TIPS FOR FURTHER INVESTIGATIONS

This section contains tips for the Student Activity Pages that will help teachers who choose to have students complete the Further Investigations.

TRICK 1 – The Missing Eight

1. Noting that $123456789 \times 9 = 1111111101$, use an analysis similar to the one used in the original trick.
2. One possibility is writing:

   $12345679 = (1234 \times 10^4) + 5679$

   Now,

   $63 \times 12345679 = (63 \times 1234 \times 10^4) + (63 \times 5679)$

   $= 777420000 + 357777$

   $= 777777777$
3. Use a calculator and inductive reasoning.
4. Use a calculator and inductive reasoning.

TRICK 2 – The Human Calculator

1. Suppose you wish the final sum to be 3581 and you write 584 as the first number on the list. After each selection of a three-digit number by the subject, choose a three-digit number such that the sum of the two is 999. If this is done three times, what is the sum of the seven numbers?
2. You might suggest selecting numbers between 10000 and 90000.
3. After the subject selects each of the first three numbers, you write the arithmetical complement of each number. Will there be a need to subtract 3 from the final sum?

TRICK 3 – Foretelling a Sum

Reference access is desirable so students may investigate *Fibonacci* and the *Fibonacci numbers*.

1. Are there any numerical restrictions on x and y for the algebraic computations performed in the explanation of the trick? If x and y were rational numbers (written as fractions or decimals), would it be easy to mentally multiply the seventh entry by 11?
2. Try several cases and it will become inductively clear that the sum is one less than the number appearing two steps below the line. For a list of twenty *Fibonacci* numbers this result can be proved by trying all cases, that is, using a proof by exhaustion. If one is familiar with proof by mathematical induction, the trick can be proved for any number of terms (greater than two) in a *Fibonacci* series.
3. Try several cases. In each case subtract your sum from the number two steps below the line. What do you observe? Try the same algebraic analysis used in this trick for some specific cases in the series.

TRICK 4 – Unveiling Even and Odd

1. In Case 1 assume the first person selects the odd number (call it x) and the second person selects the even number (call it y). Show that $2x + 3y$ is even, allowing you to predict that the second person took the even number. In a similar way consider Case 2 .
2. You might consider having the subjects divide their sum by 2 and use this quotient to predict which number each person selected.
3. Does the algebraic analysis of the trick specify the odd and even numbers used?
4. Try a method similar to that used in analyzing the original trick.
5. Do not use a fifty-cent piece and a dime for they are both represented by even numbers.
6. Suppose $x = 12$ and $y = 7$ where the prime $p = 2$ divides x. If A selects 12 and multiplies by the prime $p = 5$, the result is 60. B computes $7 \times 2 = 14$ and the sum is $60 + 14 = 74$. Since the sum is divisible by p, subject A chose x or 12. Similarly, examine what happens if A chose y.

TIPS FOR FURTHER INVESTIGATIONS

TRICK 5 – Countermoves

1. One possible set of moves is as follows:

   a. Move n from Pile 1 to Pile 3.

   b. Move n from Pile 5 to Pile 3.

   c. Remove Pile 1.

   d. Remove from Pile 3 as many counters as are in Pile 5. Remove Pile 5.

   e. Move Pile 2 to Pile 3.

   f. Remove from Pile 3 as many counters as are in Pile 4. Remove Pile 4.

   Then announce the number left on the table.

2. At the end, you will find that there are 2 in Pile 1 and none in Pile 2. Make a table for the two Piles, and follow the trick step-by-step.

3. R stands for *red* and G for *green*:

   RRRRGGGGGRRGRRRGRGGRRGGGRGGRRG

TRICK 6 – Predicting the Hour

1. $12 - (15 - (x + 3)) = x$. You should be pointing to 12 for the third number.

2. For a 10-hour clock use the equation:

   $10 - (20 - (x + 10)) = x$

   If you wish the subject to silently count to 20. You should be pointing to 10 for the tenth number.

TRICK 7 – Amazing Prognostication

1. There are four cases to check. In one of these cases, your prediction is 1 and the volunteer chooses 50. Then your key is $99 - 1 = 98$ and the equations below are those calculations required by the trick:

   a. $50 + 98 =$ 148

   b. 148 → ~~1~~48

   c. $48 + 1 =$ 49

   d. $50 - 49 =$ 1

   Exactly one of these four cases does not work which is why the choices were given by strict inequalities.

2. Subtract your prediction from 999 and follow the remaining steps in the original trick.

TRICK 8 – The Divining Rod

1. Since there are $9 \times 9 \times 8 = 648$ possible choices for the original three-digit number, it depends on how ambitious you are and whether you are able to enlist the help of a computer.

2. It works, if you consider the permutations with leading zeros to be two-digit numbers. Try it with the set of permutations 250, 205, 025, 052, 520, and 502.

3. If two digits match, there are only three permutations, and the sum may not be divisible by 2. Try dividing the sum by 3 and the result by the sum of the digits. If all three digits are identical, how many permutations are there?

4. Start with a four-digit number that has four distinct digits: a, b, c, and d. You should find 24 permutations and $S = 6666(a + b + c + d)$. Look for a series of arithmetic computations to convert S into 37.

TRICK 9 – Predictable Dice

1. It is important that the sum of the top numbers and the bottom numbers on any polyhedral dice used are constant, otherwise the trick will not work. Try several rolls (or simulated rolls) for a pair of dodecahedra dice. Inductively it will become clear that the sum of the four products is 169. If x represents the number on top of a *dodecahedron*, how would you express the number on the bottom? A verification of this trick can be made following the pattern of the algebraic proof used for standard dice. How many possible rolls are there for a pair of *dodecahedra* dice? Could you use a proof by exhaustion? Would you want to?

2. Again try several rolls (or simulated rolls) for *icosahedra* dice. What do you notice about the sum of the four products? Try the same technique of proof used previously. Could you use a proof by exhaustion?

3. The result is always 147 (3×49). This can be discovered inductively by trying several rolls of three dice. There are twelve products involved and the proof follows the same format as that used for two dice. How many possible rolls are there for three dice? How could you do a proof by exhaustion with a little help from your friends?

TRICK 10 – Who's Hiding the Coin?

1. If the identification number of the concealer is a two-digit number, the final computation will produce a three-digit number. The two-digit number formed by the two left-most digits, the hundred's digit followed by the ten's digit, will identify the concealer and the one's digit, the hand in which the coin is held.

2. Indeed, as the previous investigation suggests, the identification number can be any counting number. The number of digits in the final computation will be one greater than the number of digits in the largest identification number with the one's digit revealing which of the concealer's hands is used and the remaining digits revealing the identification number. One can show this algebraically using expanded notation for the identification number much like the algebraic analysis in *Trick 16–The Missing Digit.*

3. Let x equal the person's I.D. number, y equal 1 or 2 depending on the hand used, and z equals a digit from 1 to 6 identifying the type of coin. Following the Steps in the Trick, one arrives at the solution: $100x + 350 + 10y + z$. When you subtract 350 from this answer, you can identify x, y, and z.

4. Follow the algebraic analysis suggested in *Trick 4–Unveiling Even and Odd.*

TRICK 11 – So How Am I Doing?

1. For example: Step 6 – Add 5.

   Step 7 – Divide by 2.

   Step 8 – Your prediction is always 7.

2. Note that

$$\frac{(100a + 10b + c) - (100c + 10b + a)}{9} =$$

$$\frac{(99a - 99c)}{9} = 11(a - c)$$

3. Suppose N is the number selected. After casting out nines, you are left with the remainder R_1. Note that R_1 can be found by repeatedly summing the digits of N. Suppose the digit d is deleted from N. When casting out nines from the remaining digits, you will be left with remainder R_2. Thus, R_1 should be the same remainder as that found when casting out nines in $R_2 + d$. Knowing R_1 and R_2, you can predict d as follows:

   a. If $R_1 = R_2$, then a 9 was deleted

   b. If $R_1 > R_2$, then $d = R_1 - R_2$

   c. If $R_1 < R_2$, then $d = 9 - (R_2 - R_1)$

   For example, if $N = 387$, then $R_1 = 0$. If the 8 is deleted, then $R_2 = 1$. Since $R_1 = 0$, then casting out nines in $R_2 + d$ should also give a zero remainder. Therefore, $R_2 + d$ or $1 + d$ should have a remainder of 0 when casting out nines. Thus,

   $d = 9 - (R_2 - R_1) = 9 - (1 - 0) = 8$

4. Suppose the original number is $N = 1000a + 100b + 10c + d$ and the digit b is deleted. Then, $(100a + 10c + d) - (a + b + c + d) = 99a + 9c - b$. If a digit had not been deleted, this would be a multiple of 9. After casting out nines (by adding the digits), the result subtracted from the nearest multiple of 9 should tell you the deleted digit. Note that if either 0 or 9 are deleted, the answer will be 0.

TIPS FOR FURTHER INVESTIGATIONS

TRICK 12 – The Marble Transfer

1. Assume there are a lot of marbles in the bowl. Would the trick work if A originally took 2 marbles? 3? 4? 5? More than 5? To ensure that the trick will work, restate the first direction to A so he will have enough marbles to make the trick possible.
2. Since you know B has 20 marbles, any transfers or computations that follow will allow you to keep track of the number held by B.
3. Simulate the trick for several different selections by subject A. Can you use the same method of proof used in the original trick?
4. Suppose A takes n marbles. Then B takes pn marbles. After the first transfer, the number held by A and B, respectively, are $n - q$ and $pn + q$. After the second transfer, the number held by each is $(n - q) + p(n - q)$ and $pn + q - [p(n - q)]$. Show this last expression is equal to $q(1 + p)$. When A announces that he holds x marbles, show that the number of B's marbles added to x and then divided by $(1 + p)$ is the original number taken by A.
5. First transfer 10 white marbles from Bowl 1 to Bowl 2. Now let x be the number of black marbles and y the number of white marbles involved in the transfer from Bowl 2 to Bowl 1. How are x and y related? Write algebraic expressions for the number of white marbles in each bowl. Do the same for black. Compare these expressions.

TRICK 13 – A Permutation Trick

1. There are 24 permutations. The sum in expanded notation is $100(2a + 2b) + 10(4a + 4b) + (6a + 6b) + {}^1/_{10}(6a + 6b) + {}^1/_{100}(4a + 4b) + {}^1/_{1000}(2a + 2b) = 246.642(a + b)$.

   What whole numbers divide 246642?
2. The following set of instructions lead to an answer of 0: multiply by 100, divide by 11, divide by 3, divide by 37, and from this result subtract the sum of the two numbers on the cards.
3. There are 24 permutations and the sum is:

   $100(2a + 2b + 2c) + 10(4a + 4b + 4c) + (6a + 6b + 6c) + {}^1/_{10}(6a + 6b + 6c) + {}^1/_{100}(4a + 4b + 4c) + {}^1/_{1000}(2a + 2b + 2c)$

   Analyze the result as was done in the original trick.
4. Show that $100a + 10a + a$ divided by $3a$ is 37.

TRICK 14 – Permutation Prediction

1. Try the same numerical and algebraic analysis used in the original trick but with $2a$ to represent the hundred's digit if it is even and $2a + 1$ if it is odd. You'll need to divide by 12.
2. The result is nice because it yields a single-digit predictor.
3. The answer is always a constant.
4. The trick can be adapted, but because there are only two possible permutations of a two-digit number there is little magic in determining the consecutive ascending digits.
5. There are 24 permutations of a four-digit number. For each of the 6 possible starting numbers, consider the results when the sum of the permutations is divided by 12. Combine these results with an algebraic analysis. (Column sums divided by 12 are all $2a + 3$.)

TRICK 15 – Divining Two Numbers

1. If the numbers are denoted by x, y, and z, so far the computations yield $100x + 10y + 55$. Simply add the third number and subtract 55 to arrive at the three-digit number xyz which divines the chosen numbers.
2. Simply continue the steps in the first investigation.
3. If x represents the first number then the algebraic result of the directions given thus far is $100x + 55$. Simply add the second number and subtract 55 to arrive at a four-digit number whose digits are the first two-digit number followed by the second two digit number.
4. As described above, the result would be $100x + y$.

TRICK 16 – The Missing Digit

1. $N - S = (100a + 10c + d) - (a + b + c + d) = (99a + 9c) - b$. To produce a multiple of 9, you must add the digit b.

2. When the difference is a multiple of nine you would not be able to tell whether the deleted digit was a 0 or a 9.

3. If the digit 4 is deleted from the six-digit number 387241, then: $N - S = 38721 - (3 + 8 + 7 + 2 + 4 + 1) = 38696$ and $3 + 8 + 6 + 9 + 6 = 32$. You must add 4 to 32 to get the next multiple of 9.

4. Write the three-digit number as abc. Assume $a > c$. Subtract as shown. Note that the sum of the first and last digits is 9.

$$\begin{array}{rccc} & a & b & c \\ - & c & b & a \\ \hline \end{array} \qquad \begin{array}{rccc} & a-1 & b+9 & c+10 \\ - & c & b & a \\ \hline & (a-1-c) & (b+9-b) & (c+10-a) \end{array}$$

5. Since 457 leaves a remainder of 7 upon division by 9, and 47 leaves a remainder of 2, the deleted digit was $7 - 2 = 5$. Note that 590 leaves a remainder of 5 upon division by 9, as does 50 after the 9 is deleted. Could you tell whether a 0 or a 9 was deleted?

6. For a four-digit number written *abcd*, using expanded notation one has:

   $1000a + 100b + 10c + d + 8(a + b + c + d)$

   $= 1008a + 108b + 18c + 9d$

   $= 9(112a + 12b + 2c + d)$

TRICK 17 – Think of a Number

1. (a) Think of x and the result is $10x + 3$.

   (b) Think of x and the result is $100x + 320$.

   (c) Think of x and the result is $2x - 1$.

TRICK 18 – Magical Number Line

1. Suppose the number chosen can be written as ab where $a > b$. Then we have:

$$\begin{array}{cc} \text{ten's} & \text{unit's} \\ a & b \\ -b & a \\ \hline \end{array} \quad \text{which gives} \quad \begin{array}{cc} \text{ten's} & \text{unit's} \\ (a-1) & (b+10) \\ -b & a \\ \hline (a-1-b) & (b+10-a) \end{array}$$

   Note that the sum of the digits in the difference is: $(a - 1 - b) + (b + 10 - a) = 9$. So if you know one digit in the difference you know the other.

2. Suppose 87 is chosen. $87^2 = 7569$ and $78^2 = 6084$.

   $7569 - 6094 = 1485$ and $1 + 4 + 8 + 5 = 18$.

3. The sum of the digits of any difference must be a multiple of nine. This is the basis of all the tricks above and casting out nines. Knowing all but one of the digits allows you to determine the missing digit, simply add the given digits and determine what digit brings this sum up to a multiple of nine. This digit will be unique.

   Notice that knowing only one digit is enough. For example, knowing the sum of the digits ends in 7 only tells you that the sum of the digits could be 27, 117 or any of the infinitely many other multiples of 9 which end in 7.

TRICK 19 – Divining From a Table

1. Since $2^8 = 256$ and $2^9 = 512$, it would take nine cards.

2. The base two numeral 11111111 is 255 in base ten. Note that 255 is equal to $2^8 - 1 = 256 - 1$.

3. Another example is 37 which is 100 101 in binary.

4. Since the division, $50 \div 2 = 25$, leaves a zero remainder, there is a 0 in the unit's place in the binary code for 50. Since $25 \div 2 = 12$ with a remainder of 1, the twenty-five groups of two can be regrouped into twelve groups of four with one 2 left over. This 2 is represented by a 1 in the two's column of the binary code. Follow this reasoning as you continue the divisions by 2.

5. The first twenty counting numbers appear in their decimal and ternary codes in the table below.

   Since each place value in the ternary system may use the digits 0, 1, or 2, it is necessary to identify which numbers

Decimal	Ternary
1	1
2	2
3	10
4	11
5	12
6	20
7	21
8	22
9	100
10	101
11	102
12	110
13	111
14	112
15	120
16	121
17	122
18	200
19	201
20	202

   have a 1 and which a 2 in any place value location. Consider the following scheme for making cards where two cards are used for each place value location in the ternary code.

   The first numbers appearing on the cards above would act as a key as was done in the original trick. The numbers 1–20 can be predicted using these six cards.

Place value	Card number	First 20 counting numbers
units	1	1, 4, 7, 10, 13, 16, 19
units	2	2, 5, 8, 11, 14, 17, 20
threes	3	3, 4, 5, 12, 13, 14
threes	4	6, 7, 8, 15, 16, 17
nines	5	9, 10, 11, 12, 13, 14, 15, 16, 17
nines	6	18, 19, 20

TRICK 20 – The Domino Chain

1. Starting with a 3–3 domino one might reason as follows. If a 3–3 domino starts the chain, it must be followed with a 3–x domino (with $x \neq 3$). This leaves five dominoes on which a 3 appears on only one end. But, as they are used in the chain, they must be paired, which leaves one 3–y domino (with $y \neq 3$) that is not paired with a 3. This remaining domino must be paired with a y and this leaves the 3 on the other end of the chain.

2. If you conceal a double, say 3–3, this leaves six dominoes with a 3 on only one end. These six can be paired in the completed chain so the chain could end with a number that is not a 3.

3. No. It takes a minimum of 3 chains.

4. To verify this, it is useful to know the formula for summing the first n counting numbers:

$$1 + 2 + 3 + ... + n = \frac{n(n+1)}{2}$$

Now, consider the set of double-six dominoes arranged as follows:

0–0

0–1 1–1

0–2 1–2 2–2

0–3 1–3 2–3 3–3

0–4 1–4 2–4 3–4 4–4

0–5 1–5 2–5 3–5 4–5 5–5

0–6 1–6 2–6 3–6 4–6 5–6 6–6

The number of dominoes in the seven rows is:

$1 + 2 + 3 + 4 + 5 + 6 + 7$

Using the formula for summing consecutive counting numbers we conclude:

$$1 + 2 + 3 + 4 + 5 + 6 + 7 = \frac{7(7 + 1)}{2} =$$

$$\frac{(6 + 1)(6 + 2)}{2} = \frac{56}{2} = 28$$

This argument can be generalized for a double-n set of dominoes.

For Further Reading

Figuring
by Shakuntala Devi, **Harper & Row Publishers**, 1977

536 Puzzles & Curious Problems
by Henry Ernest Dudeney, **Charles Scribner's Sons**, 1967

Fun with Mathematics
by Jerome S. Meyer, A Premier Book, **Fawcett World Library**, 1957

Magic House of Numbers
by Irving Adler, **The John Day Company**, 1957

Magic Tricks, Card Shuffling, and Dynamic Computer Memory
by S.B. Morris, **Mathematical Association of America**, 1998

Martin Gardner's New Mathematical Diversions from Scientific American
by Martin Gardner, **Simon and Schuster**, 1966

Math Horizons, including
"Top 5 reasons to like mathematical card tricks,"
"Conversation with Jason Latimer, math major and magician," and
"Unshuffling for the imperfect mathematician."
Mathematical Association of America, February 2004

Mathemagic
by Royal Vale Heath, **Dover Publications**, 1953

Mathemagic
by Raymond Blum, **Sterling Publishing Company**, 1992

Mathematical Card Tricks
www.ams.org/new-in-math/cover/mulcahy1.html

Mathematical Magic
by William Simon, **Dover Publications**, 1964

Mathematical Puzzles & Diversions
by Martin Gardner, **Simon and Schuster**, 1959

Mathematical Recreations and Essays,
by W.W. Rouse Ball, **Macmillian and Company**, 1931

Mathematics, Magic, and Mystery
by Martin Gardner, **Dover Publications**, 1956

Mathemagic: Magic, Puzzles, and Games with Numbers
by R.V. Heath, **Dover Publications**, 1953

My Best Puzzles in Logical Reasoning
by Hubert Phillips, **Dover Publications**, 1961

Puzzles & Posers
by Philip Haber, **Peter Pauper Press**, 1963